grass roots

How to Fight Nuclear Power and Win!

grass roots

AN ANTI-NUKE SOURCE BOOK

EDITED BY FRED WILCOX

 THE CROSSING PRESS / TRUMANSBURG, NEW YORK 14886

To Phoebe, Gilea, Danica and Brendan. And to Native Americans whose courage and wisdom I greatly admire.

Acknowledgments:

Thanks to Anna Gyorgy for her help and advice, and to Dan Burgevin and Jerry Przygocki for their cartoons, and to all people throughout the world who are dedicating their lives to putting an end to nuclear power and the arms race.

Library of Congress Cataloging in Publication Data
Main entry under title:

Grass roots.

  Bibliography: p.
  1. Atomic energy policy--United States--Citizen participation--Handbook, manuals, etc. 2. Politics, Practical--Handbooks, manuals, etc. I. Wilcox, Fred.
HD9698.U52G67   333.79'24   80-11762

ISBN 0-89594-032-9
ISBN 0-89594-031-0 (pbk.)

PREFACE

Following the near meltdown at Three Mile Island a nonpartisan committee was established to investigate the accident and make recommendations to the President. The Committee concluded that the Nuclear Regulatory Commission had been remiss in handling the crises at TMI and that the Commission should be abolished, but the investigators were unable to reach a consensus for declaring a moratorium on nuclear power. President Carter did not abolish the NRC, nor did the commissioners resign en masse, though they did declare a six month moratorium on issuing construction or operating permits for nuclear reactors. Shortly after the NRC's decision Congress voted against a moratorium, Mr. Carter once again expressed his support for a nuclear future, and Stuart Udall, considered an ally by some environmentalists, announced his reluctant endorsement of nuclear power.

Some opponents of nuclear power have suggested that Three Mile Island destroyed public confidence in nuclear power and frightened investors away in such large numbers that financing for new plants is not easy to secure. Certainly the accident at TMI, combined with pressure from the anti-nuclear movement, has frustrated the expansionist dreams of the nuclear industry in some areas of the country. But to assume that the industry is finished would be a very strategic mistake. Nuclear power may have been in abeyance for a time due to the shock of TMI, but there is no evidence that it is dead, finished, or that the industry lacks the will to continue its relentless destruction of the environment.

Today there are still close to sixty-five operating plants in the United States and ninety more in various stages of planning or construction. The waste is still piling up around reactors. Trucks, trains and even planes loaded with radioactive materials are still passing through and over our communities. Radioactive isotopes are still being released into the air we breathe and the water we drink. The government has plans to reopen the radioactive swamp at West Valley, New York, strip mine the Black Hills of South Dakota for their uranium deposits, and continue pouring tax payers' money into research on the breeder reactor. Nuclear Power is not dead. And it will be finished only when every plant is closed and the waste disposed of in a manner which will protect future generations from our mistakes. Until then, the struggle against nuclear madness must continue.

Fred Wilcox
Trumansburg, New York

CONTENTS

INTRODUCTION

One fall afternoon in 1977, I was watching a professional football game with a friend who worked for a large construction company. But the game was dull so we turned down the sound and my friend, who loved to talk almost as much as he loved football, told me about the latest American boondoggle, nuclear power.

Having worked on high rise construction I wasn't surprised to hear about shoddy work--manship, corners cut to meet deadlines, heavy drinking on the job and the profits to be made using the tricks of the trade. But as I listened, I felt anxious. I wanted to argue that there was a qualitative difference between constructing high rise apartment buildings and atomic reactors. I was irritated with my friend's blithe acceptance of so much corruption, yet reluctant to argue that what he was doing was dangerous.

Driving home that night I felt frustrated. Why hadn't I told him about an article I had read just a few days before where I learned that no private insurance company would in--sure the nuclear power industry because the risks were simply too great? Why hadn't I challenged him? By the time I reached my driveway I understood why. The United States is the most technological society in the world, and I was not a scientist, biochemist or nuclear physicist. How could I, a mere citizen, question something as complex as nu-clear power?

I spent several days thinking about what my friend had told me and the more I thought about it, the more determined I became to find out all I could about nuclear power. And so I began what would turn into a two year study of atomic power, a study which would take me to symposiums on radioactive waste, debates on and demonstrations against nuclear power, interviews with Clamshell activists, professors of biochemistry, family therapists, doctors, and even one guru. I wanted to believe that nuclear power could be beneficial to our world in some way. But the more I read and the more experts I heard defending the atomic industry, the more apparent it became that nuclear power poses a terrible threat to our world.

It was not always easy to see through the subterfuge and statistical acrobatics of pro-nuke scientists. For example, when I heard Mike McCormack, the only scientist in Con-gress, state at a symposium on radioactive waste that, "The situation at Hanford poses no threat to the biosphere," I said, 'Well, maybe Congressman McCormack knows some-thing I don't.' But reading John Berger's Nuclear Power: The Unviable Option *convinced me that not only was Mike McCormack misinformed, he was dangerously uninformed. At a debate on nuclear power I would hear Nobel Laureate Hans Bethe say that "Working on your roof is more dangerous than living near a nuclear plant." 'Remember,' a little voice warned, 'he's a professor and you're not.' But after reading* John Gofman's and Arthur Tamplin's Poisoned Power *I understood why people laughed and even booed when Mr. Bethe made his ridiculous statement.*

Often during my research I would be reminded of something Native Americans have been trying to tell us for at least two hundred years: "If you kill all the buffalo," they said, "there will be nothing left to eat. If you destroy all the water, there will be nothing to drink. If you cut down all the trees, there will be no shade. Without water, food, or shade you will die." Simple statements. Simple wisdom.

Everywhere I've traveled in my efforts to compile information for this book I've met people who are not intimidated by the nuclear industry. I've talked with farmers, truck drivers, housewives, school teachers and carpenters who have taken the time to examine the issue of nuclear power and have decided that the risks are simply too great and the advantages too few. Before Three Mile Island, many of the people with whom I talked believed that only an engineer or scientist could understand nuclear power. They had approved of nuclear power because they didn't feel they were qualified to disapprove. Today, they are actively involved in the struggle to stop this menace.

GRASS ROOTS is for people who have concluded that nuclear power is the ultimate insanity. It is for the growing number of citizens who believe that nuclear power must be stopped before it destroys our planet.

Though I realize there is disagreement within the anti-nuclear movement on just how to accomplish this task, I believe that each of the articles in this collection is vital to the struggle. For some people, intervening in a government hearing or going to a stockholders meeting might appear a more direct approach than starting a citizen's initiative. Others may find challenging the nuclear industry's media blitz more rewarding than attending a civil disobedience demonstration. During the process of putting this book together, I found there are many ways to tackle a giant.

Whether you choose to work within or outside the system, alone or with a group, GRASS ROOTS should provide you with useful information. But even while we work to protect our towns from becoming dumping grounds for radioactive waste, our highways and railroads from carrying deadly cargoes, and our countryside from turning into a radioactive wasteland, we should be wary of our own victories. Although our accomplishments on a grass roots level are vital to the overall struggle, it is clear that only when nuclear power and nuclear arms are banned from the entire planet can we feel secure. Only then can we be certain that our rivers, lakes, air, our bodies and our childrens' bodies will be free from radioactive contamination.

Today, people are joining hands all over the world, sharing their strength and ideas to defeat the nuclear monster. Though the struggle may be long and often difficult, together we will win.

*Fred Wilcox
Trumansburg, New York*

HOW TO FORM A GRASS ROOTS ORGANIZATION

The nuclear industry is well financed, powerful, and absolutely determined to survive. If we are to defeat nuclear power we must work together. In some areas of the country people have been active on the grass roots level for many years, while in others groups are just now being organized to oppose nuclear power. On the grass roots level people have passed transportation bans on high level radioactive waste, refused to allow their community to become a radioactive waste dump and stopped construction on a nuclear plant. Activists are working out of their homes, churches, schools and store fronts to educate the public about the dangers of the nuclear fuel cycle. There may be a long period of struggle ahead, but by combining our collective energies we will defeat nuclear power on a grass roots level.

GRASS ROOTS ACTION:
ORGANIZING FOR A CHANGE

By Anna Gyorgy, Phil Stone and Rebecca Winborn

In this article Clamshell organizers Anna Gyorgy, Phil Stone and Rebecca Winborn combine their experience as anti-nuclear activists to offer many useful ideas for organizing on the grass roots level, including: How to start a grassroots organization, How to do fundraising, and How to reach people through canvassing. This information will be extremely valuable to people who have had little or no organizing experience but who wish to work together in their own community to defeat nuclear power.

The accident at Three Mile Island in March, 1979 was not the only reason for the distinct shift of public opinion from "who cares?" to "do we have to have nuclear power?" The current sensitivity to nuclear power can be traced to the time and efforts of hundreds, even thousands, of devoted people working against atomic energy. Had they not made nuclear power an issue in communities across the United States, educating the press and public to its questionable benefits and very real dangers, we might have heard little about the accident at Metropolitan Edison's new nuclear plant. Or the true facts might have been successfully hidden. Not much attention was paid to the near meltdown at the Brown's Ferry, Alabama plant almost exactly four years earlier. And neither the Pennsylvania utility, nor the Nuclear Regulatory Commission was about to publicize the problems surrounding Three Mile Island-2. But years of work by people and groups across the country had made the media and public aware that at least some people thought nuclear power was dangerous. The reality of a severe accident made nuclear opposition understandable—and more widespread.

There are probably around 1,500 anti-nuclear citizen groups in the U.S. Because of their efforts, hundreds of thousands of concerned citizens have expressed themselves on the issue of nuclear power by writing letters, voting on nuclear questions or perhaps by going to a rally or march. The most effective citizen resistance has come in committees where people have organized and worked together to turn the tide.

Here are some practical comments, ideas, resources and just plain organizational details that we have gathered during our community work. We know many others could add from their experiences.

Getting Started

Groups get formed in different ways. Friends may form an "affinity group," meeting occasionally and coordinating actions, information and activities with other similar groups.

Being part of an affinity group is a good way to work closely with others on a project or in an action. These groups can grow organically, adding new people and splitting into two groups when the numbers get too big. Or people may choose to start a larger community group that will meet regularly and work on several issues or one major campaign. Since 1978-9, many new women's safe energy/anti-nuke groups have formed.

Some groups meet in members' homes, others find a free meeting place in a church or public building. Regular meeting times help people plan ahead. A group's meetings reflect its own priorities and internal dynamics. Many groups have a rotating facilitator for meetings instead of the same strong chairperson. Meetings start with introductions of all present (going around), and newcomers are welcomed and given a brief overview of the group's philosophy and strategy. New members are asked about their background and interests and what they hope to gain from and contribute to the organization.

Any new group should find out who its friends are. Get in touch with any energy, environmental and other sympathetic groups in your area. Find out what they have done, are doing, plan to do. Get their newsletters and invite their members to speak at your meetings. Groups in an area can support each other even while following quite different approaches. The efforts of legal interveners, for example, will be strengthened by carefully organized nonviolent direct actions and public education campaigns. And vice-versa.

Subscribe to major national energy, anti-nuclear publications. Although individual subscriptions vary from $3 to $12 per year, most periodicals offer bulk-rates to groups. Learning from other groups and keeping up with current nuclear/energy/utility news are important to any local group. It's usually how we find out what others are doing. Magazines and book collections let members and friends educate themselves (try a lending library) and provide material for local newsletters.

Once you've got a group, you'll want to publicize your existence and begin to reach out and educate your community about nuclear power and alternatives.

The focus your group chooses will determine the kinds of things you do. Groups that want to take legal action against nuclear plants by intervening will find a network of people, groups and lawyers that have been doing this for some time. But most groups, even if they are involved in some kind of legal intervention or court case, also have public education and building citizen political pressure as major priorities. So do groups that are more focused on direct action. Public education is a must. Grass roots groups across the country are spreading the facts community by community, countering years of government and utility misinformation and "hard sell" of nuclear power. The next few years will be a crucial watershed in U.S. energy policy.

Public education goes along with self-education. Study groups around the issue are great. It is important to read books and magazines. But don't wait until you are an expert to speak to others, to write your first letter or to go to a demonstration. Citizen voices in massive numbers—now—will win this fight.

Your group's focus will be determined both by the problems that face your area (nuclear plant construction; transportation of wastes; rate increases; a government munitions depot; danger from nuclear facilities some distance away; etc.) and by the individual interests of your group's members. Even areas far from actual nuclear plants are affected by the nuclear program. A major accident can harm people hundreds of miles away, and many of us are paying for nuclear-generated electricity. Groups far from "target" facilities can still oppose government and industry funding of nuclear power, and push for conservation and use of community-controlled, safe, renewable energy sources.

Whatever your group's focus, there are some very basic ways that you can get the word out. Your events should also attract new members to your group and raise some of the funds you will need to do your work.

Simple, direct ways to educate

1. Use your newspapers. Urge members and friends to write letters to the editor of your local paper about the problems —and solutions—as they see them. This creates a sense of controversy and urgency around the issue in your community.

2. A regular presence in your community—either with a weekly vigil or a regular literature table and displays—will give visibility to your work and to the nuclear issue. Vigils offer people a way to publicly oppose nuclear power without having to journey far, risk arrest, etc. Many people feel that standing up for something in their community, in full view of their neighbors, is the strongest, most effective thing they can do. What if one hour weekly Sunday vigils happened in thousands of communities across the country?

The "literature table" is the shopping center of the safe energy movement. There are attractive and informative articles and leaflets, pamphlets, magazines and books about all aspects of the nuclear problem—and energy alternatives. Literature tables can be combined with bake sales to raise money and attract people. There are now many good sources of printed materials, buttons and bumperstickers on nuclear power, conservation and alternative energy sources, nuclear weapons, and more.

3. Show a film. There are a number of excellent audio-visual resources available to community and school groups. Films, slide shows and video tapes bring home the problems of nuclear power, tell the stories of resistance to nuclear development from California to New Hampshire to Europe, show the dangers and costs of nuclear arms development, and describe the promise of conservation and alternatives.

Check the listing of audio-visual resources and then ask around for sources of films in your area. Independent film distributors often give special rates to community groups. Some university or institutional libraries either offer energy films or will consider buying ones you recommend. Larger organizations may own a copy of popular films. Take the time to investigate types and costs of these resources.

Have a speaker with your film. You don't have to have a well-known person, although sometimes you might want to. Several group members can give short presentations about the issue and what your group and others are doing.

4. Hold a workshop. Somewhere between an evening film showing and a full-blown energy fair is a weekend day or afternoon workshop. This event can be useful and even fun. Basically it brings together a variety of people and resources on a particular subject. You might start off with two or three short presentations and then break up into smaller groups for discussion. Make books and articles on the subject available for perusal or sale; hang posters; display models or exhibits. Show a film or slide show. Try to find local experts—people who have successfully done solar building; a doctor who can explain what is known about the effects of low-level radiation. Conservation workshops can explore things that can be done at home and in the community. Offer conservation awards and booby prizes. Raffle off tubes of caulking or a year's supply of powdered kelp (this seaweed blocks up-take of radioactive iodine). . .

5. The utilities go into our schools on a regular basis. So should we, to counter their pro-nuclear, pro-energy consumption propaganda with interesting, fact-filled school programs. (See Advice to teachers.)

Except for school classes, where you have a "captured audience," the success of most of your educational efforts will depend on getting the word out to your potential audience. A large turn-out does not necessarily ensure a good event, but it would be a shame for you to go to the effort of assembling resources for your community and then not attract those interested.

So advertise your meetings and events widely, hanging posters in local stores and gathering places. Give announcements to area radio and television stations, and distribute fliers. Personal contact is most effective, so tell your family, neighbors, friends and co-workers about what you're doing and invite them to join.

Reaching People: Organization Tools

The Office

An office is the home for organizational tools. More public than someone's house, an office allows people open access to your information, ideas and point of view. An office is a resource center, a meeting place, a space where volunteers can stop by and help out. It's a place where anyone can find out "what's happening," or take a look at national and out-of-region publications. Offices do not require big budgets. Sometimes you can get free space with another organization or in a socially-conscious church. Or you can pay to share space. Well run offices should bring in money. You'll get donations and money from the sale of resources. You might also ask people you know who are sympathetic but don't have time to volunteer or come to meetings to make a pledge of $5 a month or more towards the work of the office. One local Pennsylvania activist got working friends to contribute enough money for her to become a paid part-time organizer for her group. You may be surprised at the response you get. Many people won't think to offer support, but if you take the time to explain your work and needs, they'll want to help. The hardest thing is asking.

The most important part of "office management" is staffing. An office gives you visibility (even if you're on the 4th floor rear you can put a nice sign downstairs with big arrows!). But there's nothing more depressing than an office that's almost always empty. Especially when someone wants what you have to offer and finds a locked door. You do not have to be open five days a week. But do keep regular hours (three afternoons a week; 10-4, Mon.-Thurs., etc.). Post your hours clearly on the door and in all your literature, and stick to that schedule. Volunteer staffers can coordinate their work on different days by keeping a careful log and meeting together when necessary. One or more people can be in charge of re-ordering materials; another responsible for finances; another for mailing list continuity.

In some communities the office functions of distribution and information center are filled by food coops or community stores or restaurants. But a group busy on its own projects will eventually need office space, whether privately in members' homes (how many people keep boxes of files under their beds?) or in a public place.

The Mailing List

Mailing lists are important for any group. People who are interested in your activities need to hear from you. Send people on your list newsletters, announcements of upcoming events, even fundraising appeals! Begin and build up your list with a sign-up sheet at every event and meeting you sponsor. Make sure that you get people's full address, zip code (necessary for bulk-rate mailings), and telephone numbers.

The next step is to get an index card file, and mailing labels. Many stationery and office supply stores now have copying machines that print on sheets of peel-off gummed labels. Xerox centers can do it too if you bring them special "xeroxable" label sheets. The most convenient label stock is 8½x11" with thirty-three labels on a sheet.

To make a "master matrix sheet" take a piece of white paper and mark it like the label sheet. Place it with a sheet of carbon paper over a sheet of labels. Using a very dark or film ribbon, type your addresses. Save the sheets as masters from which you will photocopy more whenever you want to do a mailing. Then number your master page (the top sheet), and write that number on every label on the label sheet. (You could also mark your columns a, b, and c.) Put labels on index cards for your mailing list file (usually done in alphabetical order). When you want to change or remove a person's address label, just look up their name in the card file to see where their mailing label is.

You can use the cards to note what people have worked on, contributions made, etc., as well as any special interests or skills that might be useful to the group.

When your list approaches 500 or more names, consider having it computerized. Putting a mailing list on a computer enables you to edit and update it quickly and easily. The computer can print out your list in either alphabetical or zip code order. You can include the date the name was entered on the list and other information you find helpful. Perhaps someone in your group has access to a computer and can maintain the list. If not, commercial firms provide list maintenance services for a nominal amount. They usually charge for each label entered and printed with an editing fee for changing already entered information.

Since people move frequently, it is a good idea to "clean" your mailing list once a year, preferably in the fall. This can be done in a number of ways. You can send a first class letter (not a postcard) to everyone on your list. Write under your return address, "Return Postage Guaranteed, Address Correction Requested." By doing this, the Post Office will return to you all the letters whose intended recipient has moved and left no forwarding address. If they have moved, and filed a change of address form with the Post Office, the Post Office will inform you of their current address. A cheaper way to clean your list is to send out a third-class mailing with a tear-off coupon at the bottom of the letter for the recipient to return to you if they want to remain on your list. If they don't return the coupon, take their name off the list.

On bulk mailings. . .

The Post Office offers low rates for bulk (3rd class) mail, and even lower rates to non-profit organizations. When you have 200 names on your mailing list (the minimum number for bulk mail) consider getting a bulk rate mailing permit. You'll pay an annual fee ($40 in 1980), plus an additional fee for each piece you mail. Bulk mailings have to be sorted and bundled by zip code.

Any unincorporated group can buy a bulk-rate permit, but non-profit corporations pay lower rates. In order to qualify for non-profit rates you must be legally incorporated in your state, with an Internal Revenue Service (IRS) 501(c)3 tax-exempt status. The Post Office requires a copy of your IRS letter of status determination when you apply for non-profit organization mailing privileges.

Printed Materials

Printed materials, be they leaflets or posters announcing an event or educational leaflets about a specific issue, are one of the cheapest means of reaching lots of people.

Two printing processes especially suit our activities— mimeographing and offset printing. Finding a mimeograph machine to use is not difficult. Many churches, schools, and community organizations are willing to let others use their equipment occasionally. Offset presses are more expensive, and thus less available. If you want to go offset, inquire with commercial printers. Try to patronize union shops that will put a "bug" on their work. Also check out the resources of your larger and more established community organizations.

A mimeograph machine prints by forcing ink through holes in a stencil. The stencil can either be typed on directly, or cut on an electrostatic stencil cutting machine which can reproduce line drawings or photos. The advantages of a mimeo is that it is fast and cheap. Disadvantages are that the copy cannot have long lines like borders, or heavy solid lettering or dark areas that will cause the stencil to tear. Offset printing looks better. This way you can print borders, solid areas, half tones (photos), reverses and reductions. Its disadvantage is that it is more expensive (For excellent information on printing get *Printing It* by Clifford Burke, Book People, 2940 Seventh St., Berkeley, CA 94710, $3.)

Layout and Preparation

Ask yourself how the printed piece will be used. For leaflets, use a bold, eye-catching headline. Break up the text with graphics and keep it short. Leaflets should be small enough to fit into a pocket. 8½ x 11" paper can be folded in thirds; 8½ x 14" can be folded in half or in quarters. Leaflet text can be typed on an electric typewriter using a film ribbon or it can be typeset. Check on prices. For a major piece, it may be well worth the extra cost. Posters should be visible and readable from a distance. Avoid clutter, use borders, and have an eye-catching graphic.

Make sure you include the basics: date, time, place and directions and a contact telephone number for people to call for more information. Always put your group name and address on anything you distribute.

A good source book on layout is Nancy Brigham's "How to do Leaflets, Newsletters, and Newspapers," available from the Boston Community School, 107 South Street, Boston, MA 02111 , $1.50.

Now for paper. Your choice is important. Paper comes in standard sizes (in inches): 8½x11, 8½x14, 11x17, 17x22. Mimeo machines can only print 8½x11 or 8½x14 sizes. Paper also comes in various thicknesses or weights, ranging from 16 pound to 110 pound stock. 20 lb. paper is sufficient for printing on one side of a page. Consider buying a heavier or opaque paper for two sided printing. Use colored paper for posters and fliers as it adds another color at very little cost.

Time Lines

A time line helps you put together a variety of publicity methods to help make your campaigns or events successful.

SAMPLE TIME LINE							
4th Week	1st press release to weekly papers	Poster Design finished	Poster to Printers ————Line Up Radio Shows——————			Pick up Posters	
3rd Week	2nd press release to weekly papers	Write & send out posters ————Begin Postering—	Calendar Deadlines————————————		Pick up mailing	Get out Mailing to local list	
2nd Week	1st press release to daily papers		Tape Radio Show WXYZ		Begin phone calls to local list		Talk show WABC
Event Week	2nd press release to daily papers	Talk show	Final press release to dailies	Talk show WGHI ——-Leafleting——		Feature story in evening paper	EVENT!!

You want to remind people again and again, in as many ways as you can, that something big is happening and that they should be there.

How do you establish a presence? Posters are a good way to start. But you're going to need time to design one, and then have it printed. Where are the best places to put them up, and how many will you need? Try to have the same people poster the same areas each time. How about a mailing to people on your mailing list? Do you want to include a copy of the poster and ask them to put it up at work? Again, you'll need time to put the mailing together. How far in advance should your supporters know about the activity? Too early, they'll forget. Too late, they won't be able to make plans to attend. Do you want to approach them twice, following up the mailing with telephone calling? You could ask them if they're planning to attend, if they have any questions, or better yet, if they would like to help. How long will it take to call everyone on your list?

What about the media? There are all sorts of ways they can help you publicize an event. But remember, they've all got deadlines and they're all probably different. You've got press releases to get out to all the daily and weekly newspapers, calendar listings, public service announcements, and radio shows to line up. You could even do a series of releases, 1. an announcement of your plans, 2. giving some new details, and 3. going only to the daily papers the day before the event, giving the final program.

Money Matters

Depending on the size of your budget and group, you may want to hire a professional bookkeeper. But in most small groups a volunteer treasurer handles financial matters. As we all know, money matters can cause conflict if not handled correctly. Thus, the following suggestions.

First, every group should have a checking account and a person primarily responsible for maintaining it. It's a good idea to have two people authorized to sign checks, so you won't be stuck if one is away. Try to run all your transactions through your checking account. It then can serve as your financial record. In addition, keep a record of where all your income comes from. For example: "literature table," "meeting collection," "benefit concert," etc. Agree on your "regular" expenses. Have a process-group discussion and approval, perhaps for special or unusual expenses.

Fundraising

How can a grass roots organization maintain its impetus; how do we fund our activities? Fundraising is probably the most difficult problem small groups face. There is no simple formula for acquiring the dollars you need. The best approach is to see your fundraising events as a valuable opportunity for outreach and public education. An evening program consisting of a speaker, a film, and refreshments afterwards can easily be turned into a fundraising success.

While some group activities will never be financially self-supporting (a press release mailing to the media, for example), your public education activities should pay for themselves. A table with literature and anti-nuclear buttons, stickers, etc. should be part of all your meetings and public education events.

The success of fundraising projects will depend on the imagination of the group and the dedication of the individual members. One grass roots group, for example, decided to enter a float contest in a local fair. Though they did not win first prize, they did wear their t-shirt with the group's name and logo, and later sold a lot of shirts at the fair! Other groups have shown films, held bake sales, organized concerts, sponsored walk and bike-a-thons, made gifts to sell around holidays and placed contribution cans in the stores of sympathetic merchants in their community. Well-prepared and appetizing food is always a big money maker, particularly at rallies where it is difficult to find nourishing nibbles.

The best source of information and ideas on community group fundraising is Joan Flanagan's "The Grassroots Fundraising Book," available from: National Office, The Youth Project, 1000 Wisconsin Ave. NW, Washington, D.C. 20007. ($4.75 each plus 50¢ handling. Ask for bulk discounts.)

*See also: "Fundraising in the Public Interest" by David Grubb and David Zwick, 1976. Contact Fundraising in the Public Interest, P.O. Box 19367, Washington, D.C. 20036. $5. Make checks payable to the Center for the Study of Responsive Law.

* "101 Surefire Fund-Raising Ideas," by JoAnne Alter, October 1976 *Family Circle* magazine. Reprints are 35¢ each from the Reprint Department, Family Circle, 488 Madison Ave., N.Y., N.Y. 10022

*"Handbook of Special Events for Nonprofit Organizations: Tested Ideas for Fund Raising and Public Relations." Edwin Leibert & Bernice Sheldon. Available from Taft Products, 1000 Vermont Ave. NW, Washington, D.C. 20005. 224 pp., $13.45.

It means a lot to be supported through local fundraising, rather than by a few donors or foundations. It shows that your group represents strong community feelings and interests and "vibrates" self-sufficiency.

> "Raising money locally keeps a group accountable only to the people it serves. . . And the possibilities for local fundraising events are limited only by one's imagination." (NIRS Resource Guide)

Having said this, it must be added there that *are* situations in which support from outside the immediate community must be sought: where the local population is either unable or unwilling to fund the effort needed. Poor rural populations may just not have the money needed for major campaigns and sometimes one community's acceptance of a plant, etc. endangers the interests of a far greater area, or even another country. Although most anti-nuclear/safe energy groups draw their funding from local events—yard sales and local dinners and dances—many have sought outside funding for special projects.

Private foundations have long contributed to environmental and more recently to anti-nuclear groups. Some parts of the country have regionally-focused movement-oriented foundations. Here are the addresses and regions served by these groups:

New England   Haymarket People's Fund
120 Boylston St. Rm 707
Boston, Mass. 02116

New York State   North Star Fund
80 Fifth Ave. Rm 1203
New York, New York 10011

Pennsylvania   Bread and Roses Community Fund
1425 Walnut Street
Philadelphia, PA 19102

Pacific North-West McKenzie River Gathering
454 Willamette
Eugene, Oregon 97401
  or 19 North East Morris
Portland, Oregon 97212

San Francisco   Vanguard Public Foundation
4111 24th Street
San Francisco, CA 94114

Los Angeles   Liberty Hill Foundation
P.O. Box 1074
Venice, CA 90291

Grant Proposals

Virtually all foundations, including the "progressive" ones, will only make grants to recipients with IRS 501(c)3 tax-exempt status. In order to receive such grants directly, you must first incorporate as a non-profit organization in your respective state, and then you must apply to the IRS for tax-exempt status. Incorporating as a non-profit organization alone does not mean you are exempt from Federal taxes, or that contributors can claim their donations against their gross income for tax purposes.

One way for non-incorporated or non-tax exempt organizations to receive grants is to have a local tax-exempt organization—most churches are—receive the grant for them, and then pass the funds along. It is not too hard to find an organization willing to do this for you, but be prepared to pay up to 5% of the funds transmitted to your sponsor. Your intermediary has accounting and book-keeping expenses, as well as legal responsibilities associated with doing this for you.

Examples of projects suitable for grants include establishing a local office, hiring a local organizer, etc., but you'll find

---

BRIEF LISTING OF FOUNDATIONS WITH DEFINITE OR POTENTIAL INTEREST IN ANTI-NUCLEAR FUNDING

ARCA Foundation
100 East 85th St.
New York, NY 10028

Bydale Foundation
60 East 42nd St.
New York, NY 10017

Haymarket Foundation
2 Holyoke St.
Cambridge, MA 02138

Kaplan Fund, Inc.
Two East 34th St.
New York, NY 10016

Henry P. Kendall Foundation
One Boston Place
Boston, MA 02108

Max and Anna Levinson Foundation
95 State St.
Springfield, MA 01103

Charles Stewart Mott Foundation
500 Mott Foundation Building
Flint, MI 48502

New World Foundation
100 East 85th St.
New York, NY 10028

Rockefeller Brothers Fund
30 Rockefeller Plaza
New York, NY 10020

Shalan Foundation
2749 Hyde Street
San Francisco, CA 94109

Stern Fund
211 East 40th St.
New York, NY 10016

Abelard Foundation
1 East 53rd St., 10th Floor
New York, NY 10022

Evergreen Fund
785 County Line Road
Villanova, PA 19085

Joint Foundation
1 East 53rd St.
New York, NY 10022

Musicians United for Safe Energy (MUSE)
72 5th Avenue
New York, NY 10011

The Needmor Fund
Commodore Perry
Suite 406
Toledo, OH 43603

North Shore Unitarian Veatch Program
Plandome, NY 11030

Rockefeller Family Fund
49 West 49th Street
New York, NY 10020

Vanguard Foundation
4111 24th Street
San Francisco, CA 94114

Youth Project Foundation
1555 Connecticut Avenue, N.W.
Washington, D.C. 20036

Sources of Funds for Solar Activities, Anita Gunn. 1979. Available from Center for Renewable Resources, 1001 Connecticut Avenue, NW, 5th Floor, Washington, D.C. 20036

funding specific projects like utility campaigns, creative public education projects and the like, are usually the easiest types of grants to succeed, especially for a nuclear group. For information on new foundations being set up in Texas, Chicago and Washington D.C., contact: The Funding Exchange, 80 5th Ave. Rm. 1203, New York, N.Y. 10011.

And then there is MUSE—Musicians United for Safe Energy. Formed in 1978 by activists and popular musicians including rock and roll stars who had done many anti-nuke benefit concerts, MUSE organized five large concerts at Madison Square Garden (NYC) in September, 1979. Proceeds from the concerts, a record and future film, will go to safe energy and peace groups. For an application write: MUSE Foundation, 72 5th Ave., 2nd floor, New York, N.Y. 10011.)

A good proposal should be short and concise, no more than two or three pages long, and should include the following information:

Background of organization
Purpose of organization
Need addressed
Goal of project
Budget of project
Evaluation process
Future funding
IRS letter of tax-exempt status determination

There are very few foundations that will fund direct actions either because of the "politics" (Threatening their future ability to make tax-exempt grants), or because the amount of money requested is too small. It costs foundations just as much to administer a $5,000 grant as a $50,000 grant. If you're interested in going the grant route, read "The Bread Game," available from Glide Publications, 333 Ellis Street, San Francisco, CA 94102 ($2.95 + 50¢ for postage and handling.)

REACHING PEOPLE: CANVASSING

Canvassing is the old fashioned way to reach people—directly—by going to them door-to-door. It is used by community-base organizations, charities for fund drives and by political candidates (if they haven't gone the TV ad route). Canvassing is an important part of grass roots safe energy organizing. Canvassing involves time and truly "natural energy." When used effectively, canvassing can reach many people and sharply define an issue in a community.

Why Canvass?

You can make personal contact with people in your community, so that you are not a "faceless" political group.

You can find out what people know and don't know about nuclear power, conservation and alternative energy sources.

You can formulate and offer some real possibilities for conservation and alternative energy locally.

You can also: collect signatures for petition or ad; distribute literature; fundraise; solicit membership.

There are also good reasons for your local group to consider a canvassing project: You educate yourselves on energy issues. You create a way for people to get involved in varying ways, for varying amounts of time, in their own neighborhoods.

You are giving yourselves and others a chance to learn, practice or improve the organizing skills of:

public speaking	publicity/media work
research	strategizing, planning
writing	evaluation

You can find out your strengths and weaknesses in your community. Who are your supporters? Where do you need to do the most work?

Types of canvassing

There are several ways to canvass depending on what your goals are and the amount of people-power you can muster. Here are some general types of canvasses beginning with the most basic form, the Literature Drop, and progressing to the most complex, a multi-purpose canvass.

literature drop This means what it says, simply dropping off literature door-to-door. It is illegal to put literature into mail boxes, so use the door or mat to keep it from blowing away. The advantages of this technique:
* it is simple to organize
* it gets literature into homes
* you can cover a neighborhood quickly
* canvassers need not be trained or particularly well-versed on the subject

Disadvantages include:
- absence of personal contact
- no dialogue
- information or opinions cannot be collected
- it is the least inspiring type of canvassing

knock and drop The purpose of "knocking and dropping" is to allow for *brief* personal contact at each door and to dispense information. Don't get into lengthy discussions. Dress nicely. You only have a few seconds to make an impression.

petition drive This is an effective way to identify your supporters. You can also use collected signatures to increase your mailing list. Decide in advance how the petitions will be used to get the best publicity. The advantages of a petition drive:
* it is conducive to dialogue
* you can find out opinions in your community
* it requires an active response from supporters

Disadvantages:
- it needs more organizing than the above-mentioned types
- it is unpleasant to do in the winter
- it is not a great organization builder
- it is not very imaginative

ad canvass Utilities and multinational energy corporations advertise heavily. Once in a while your group might want to counter their fancy salespitches with an anti-nuclear ad. You may focus on a particular group (as with the "Concerned Doctors of Santa Barbara" ad), canvass a neighborhood, or do a phone canvass of potential supporters and pledges. Ask people to support the ad by allowing use of their name, and giving a donation.

multi-purpose canvassing Several goals can be accomplished within this more complex canvass. You can do a combination of the following:

* educate canvassers on issues and skills
* solicit new membership
* collect information, opinions from your community
* energize group
* fund raise

But there are considerations. For a multi-purpose canvass
- takes time for organization and follow through
- requires skilled and dedicated leadership
- can be expensive
- requires continuity; it's a long-term project
- requires committed canvassers

Before launching any canvassing effort, a group must decide if canvassing is in fact the right way to get the job done. To answer this question, "the job" needs to be CLEARLY defined. Here are some basic issues which need to be seriously considered in the early planning stages of a canvass. It may take several meetings to hash out these issues in your group, but don't get discouraged. Time taken for thorough planning is time well spent, because you will know where you are going and why, throughout the entire process.

* What is your goal? What do you ultimately want to accomplish? If you have several goals, rank them.

If canvassing seems like the best way to achieve your desired goal, which type of canvassing best suits your needs?

* How much energy can you count on for organizing and carrying out the canvass? Scale your efforts to the number of reliable volunteers available.

* What are your time constraints? Construct a timeline to keep the project moving along with the end in sight.

* What type of literature do you want to distribute? If you use an original piece, who will do the research, lay-out, printing, etc.?

* How much money do you have to spend on the canvass? What is the predicted costs of the literature, phone, mailing, etc.? If you plan on fundraising as part of the canvass, how much can you realistically expect to raise?

* How will you recruit canvassers and how many will be needed to accomplish your goal? What type of training, if any, will you provide for canvassers?

* How will you sustain the community structure that has been created by organizing a canvass network?

Canvassers and their tools

Once you decide on your priorities, your potential budget, the form and materials your canvass will use, you need to recruit and equip canvassers.

Canvassers can be found by asking friends and supporters, contacting college and high school students. You can use posters, newspaper ads or articles and direct mailings. Ask people to give a day or several hours to canvass. You may train or prepare canvassers either before they go out, or you may set up short evening sessions in each area you want to canvass. Or all canvassers can be invited to a central location for one big meeting. Canvasser training should:

* educate canvassers on the issue and in canvassing skills
* motivate canvassers and dispel any anxieties about "going public"
* organize and coordinate actual canvassing: distribute canvassing materials, select areas to be canvassed by each volunteer

Canvasser training "tools" may include:

* a fact sheet on the issues (for canvasser education/background)
* an information sheet like "some helpful hints for canvassers"
*house response forms (for easily and effectively recording information)
* maps of the areas to be canvassed. Canvassers identify their area and mark where they've been.

organizing the canvass

There may be books on this...but here is a rather straightforward way to organize a canvass. After your group has discussed the "why's" of canvassing and you have talked about and set the goals, those with the most interest, time and energy can form the canvass committee. There are a lot of things to keep track of, more than one or two people (unless they work full-time) can handle. You have to decide on the exact materials canvassers will carry and distribute; how money will be collected; how information will be gathered, collected, analyzed and evaluated. You have to see the effort from beginning. . . to end.

If you plan to ask for contributions, check with local government officials (selectpeople or the mayor's office, etc.) to see if a permit is required. You may also have to check on legislation with your state attorney general's office.

Naturally your canvass organization will be different in urban, suburban or rural areas. But the tasks remain essentially the same. Your coordinating committee will have to find town or precinct coordinators, people who will help find and coordinate the canvassers. In rural areas, the canvass might be organized by town, rather than by precinct. Basically, canvassers would be trained by canvass committee members and/or town/precinct coordinators. They would work with and check back to the neighborhood, town or precinct coordinator. A coordinator would report back to the group's canvass committee. That committee would itemize and deposit funds, organize the information coming in, and eventually analyze it and present it to the entire group for discussion and future action. It's a big job, but one that can have a broad impact on your community and the issues. But don't forget—

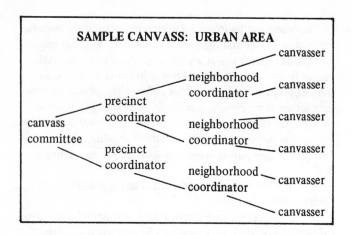

SAMPLE CANVASS: URBAN AREA

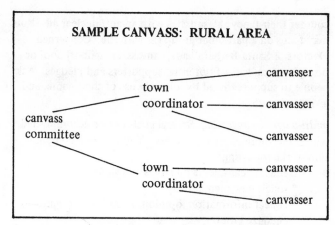

SAMPLE CANVASS: RURAL AREA

evaluation time

Just as you took time to plan the canvass, take some to reflect on it (as we all should on any action our group does) . . .Were you successful in achieving your goal? What worked well? Where were your weaknesses? How can these lessons be used to strengthen your group?. . . Where will you go from here?

one group's story

In the summer of 1977 the Franklin County Alternative Energy Coalition (AEC) planned a canvass of the 26 towns in our county in rural western Mass. We undertook the canvass as part of preparations for construction permit hearings for the twin Montague nuclear plants which we'd formed to fight back in 1974. The hearings were to begin in the late winter—early spring, and we wanted to be ready. Lawyers and many volunteer law, economics and other students were hard at work on a "citizen's intervention" that would challenge the utility in the hearings. But we wanted to have strong public pressure inside the hearings and out. So we thought a canvass would be a good way to alert people of the up-coming hearings and if possible raise the money we needed to mount even a low-cost intervention.

Our canvass committee of 6 people started meeting in July. From July to September we organized, finding town coordinators and giving them our mailing list to use in the search for canvassers. We prepared literature and canvasser preparation materials. Each canvass committee member worked with 4-5 town coordinators. Two people took responsibility for training canvassers, holding 6 sessions throughout the county, over a 6 week period.

We started canvassing in mid-September, and continued through the end of October. Every weekend one or more towns were canvassed. Around 150 people went door-to-door in their neighborhood. We added several hundred new members to our group, and distributed literature to between 3,000-5,000 households. We brought in $3,000, about half of of which went for printing and other canvass costs. But shortly after we finished, the utility delayed both construction plans and the hearings—for four years. This meant that the momentum the canvass had built up was largely lost. How-

ever, the citizen involvement and education that the canvass created was certainly important, even if it did not lead to action around the hearings. We saw the hearing delay as a good move for us. With costs rising and need declining, any delay made nuclear construction less likely in the long run.

We learned some important lessons in our canvassing effort. First of all, our grand plan was too ambitious. We asked people for too much: for a contribution to the intervention effort, for a pledge, a contribution, a membership. We had too many pieces of literature: on the plant and nuclear dangers/costs; on the intervention specifically; on solar alternatives; on economics. In the future, we would stick to two pieces, not five or six. We found that our very carefully organized town-by-town, street-by-street, house-by-house approach did not always meet local people's needs. In very rural areas where houses are located far apart, people held home meetings instead of going door-to-door. They invited their neighbors over and showed a special slide show on the project (which we had made for canvasser training), and distributed literature, signed up members, etc. We learned that structure is valuable when it is appropriate, but flexibility is crucial. We had to be responsive to community and canvasser ideas and different approaches.

We made a few basic mistakes. One was over-printing. We still have our wonderful leaflets on hand, now somewhat out-of-date. And we over-extended ourselves. By the end of October the canvass committee was so tired that we never analyzed the canvass responses adequately. Since the hearings were cancelled, we had an excuse for not carefully evaluating response to the issue town by town. But we could have done more with our information. Be careful of "burn-out" in any organizing effort. Citizen action needs sustained energy as well as those big bursts where hundreds act together.

If we had it to do again, we would simplify the canvass by concentrating on one issue and a membership drive. We still run into people who remember their day of canvassing as an important action for them. And we know that although this was our most ambitious effort, there were positive results. Over the years, our group has done door-to-door work regularly, on two meeting votes and referenda questions. We have always seen that personal contact is an important part of organizing to stop nuclear power.

A RESOURCE BIBLIOGRAPHY

BOOKS

Berger, John J., *Nuclear Power: The Unviable Option,* Ramparts Press, Inc., Palo Alto, Ca., 1976. $4.50.

The Book Publishing Company, *Shutdown: Nuclear Power on Trial as Experts Testify in Federal Court,* 1979, 156 Drakes Lane, Summertown, Tennessee 38483. $4.95.

Caldicott, Helen, *Nuclear Madness,* Autumn Press, Inc., 25 Dwight St., Brookline, Mass., 02146, 1978. $3.95.

Council on Economic Priorities, *Jobs and Energy: The Employment and Economic Impacts of Nuclear Power, Conservation and Other Energy Options,* New York, 1979.

Environmental Action Foundation, *Countdown to a Nuclear Moratorium,* 724 Dupont Circle Bldg., Washington DC, 20036. $3.00.

Environmental Policy Institute, *Plutonium and the Workplace: An Assessment of Health and Safety Procedures for Workers at the Kerr-McGee FFTF Plutonium Fuel Fabrication Facility, Crescent, Oklahoma,* 317 Pennsylvania Ave. SE, Washington DC, 20003. $4.00.

Environmental Policy Institute, *Radiation Standards and Public Health: Proceedings of the Second Congressional Seminar on Low-Level Ionizing Radiation,* 317 Pennsylvania Ave. SE, Washington DC, 20003. $4.00.

Faulkner, Peter, *The Silent Bomb,* Environmental Action Reprint Service, Box 545, LaVeta, Colo., 81055. $3.95.

Fuller, John G., *We Almost Lost Detroit,* Reader's Digest Press, 1975.

Gofman, John W., *"Irrevy" an Irreverent, Illustrated View of Nuclear Power,* Committee for Nuclear Responsibility, 2140 Taylor (rm. 1101), San Francisco, CA. 94133. $3.95

Gofman, John W., and Tamplin, Arthur R., *Poisoned Power: The Case Against Nuclear Power Plants Before and After Three Mile Island,* Rodale Press, Emmaus, PA., 1971. $9.95. New edition, 1979.

Grossman, Richard, and Daneker, Gail, *Energy, Jobs and the Economy,* Alyson Publications, Inc., Box 590, Central Square, Cambridge, Mass., 02139, 1979. $3.45.

Gyorgy, Anna and Friends, *No Nukes: Everyone's Guide to Nuclear Power,* South End Press, Box 68, Astor Station, Boston, Mass., 02123, 1979. $8.00. (40% discount to safe energy groups)

Hayes, Dennis, *Rays of Hope: The Transition to a Post-Petroleum World. Nuclear Power: The Fifth Horseman,* Environmental Action Reprint Service, Box 545, LaVeta, Colo. 81055, 1976. $2.00.

Honiker vs. Hendrie, A Lawsuit to End Atomic Power, The Book Publishing Co., 156 Drakes Lane, Summertown, TN 38483. $5.00.

Inglis, David Rittenhouse, *Nuclear Energy: Its Physics and Its Social Challenge,* Reading, Mass.: Addison-Wesley Publishing Co.

Leaf Research Team and Huver, Charles, M.D., *Toward a Realistic Fission Dose Estimate: Methodology and Case Study,* Leaf, Inc., Rt. 6, Box 262, Stevens Point, Wisc., 54481. $5.00.

Lovins, Amory, *Soft Energy Paths: Toward A Durable Peace,* Harper & Row edition, 231 pps. $3.95. Ballinger edition, paper, 251 pps. $6.95 from Friends of the Earth, 124 Spear St., San Francisco, CA. 94105.

Miller, Jack, *A Primer on Nuclear Power,* Anvil Press, Box 37, Millville, Minn. 55957. $2.50.

Morgan, Richard, *Nuclear Power: The Bargain We Can't Afford,* Environmental Action Foundation Reprint Service, Box 545, LaVeta, Colo. 81055. $3.50.

Nader, Ralph, and Abbotts, John, *The Menace of Atomic Energy,* W.W. Norton & Co., Inc., 500 Fifth Ave., New York, N.Y. 10036. $10.95.

Nash, Hugh, editor, *The Energy Controversy: Soft Path Questions and Answers by Amory Lovins and his Critics,* Friends of the Earth Books, 124 Spear St., San Francisco, CA. 94105. Cloth $12.50. Paper $6.95.

Nero, Anthony V., *A Guidebook to Nuclear Reactors,* University of California Press, 1979. $9.95.

Okagaki, Alan, *Country Energy Plan Guidebook: Creating a Renewable Energy Plan,* Institute for Ecological Policies, 9208 Christopher St., Fairfax, Virginia 22031. 1979. $7.50.

Olson, McKinley C., *Unacceptable Risk: The Nuclear Power Controversy,* Bantam Books, 666 Fifth Ave., New York, N.Y., 10019. 1976. $2.25.

Patterson, Walter C., *Nuclear Power,* Environmental Action Reprint Service, Box 545, LaVeta, Colo. 81055. $3.50.

Pollard, Robert, *The Nugget File,* Union of Concerned Scientists, 1208 Massachusetts Ave., Cambridge, Mass., 02138. 1979. $4.95.

Steinhart, John, *Path Way to Energy Sufficiency: The 2050 Study,* Friends of the Earth Books, 124 Spear St., San Francisco, CA. 94105. Paper $4.95.

Tye, Lawrence S., *Looking But Not Seeing,* Union of Concerned Scientists, 1208 Massachusetts Ave., Cambridge, Mass., 02138. 1979. $3.50.

Union of Concerned Scientists, *The Nuclear Fuel Cycle,* Environmental Action Reprint Service, Box 545, LaVeta, Colo., 81055. 1975. $4.95.

Young, Louise B., *Power Over People,* Oxford University Press, Box 900, 16-00 Pollitt Drive, Fair Lawn, N.J., 07410. 1973. Paper $2.95. Cloth $7.50.

For an excellent, detailed energy bibliography, write to Friends of the Earth, 124 Spear St., San Francisco, CA., 94105.

REGIONAL NEWSLETTERS

Coalition for the Environment
Missourians for Safe Energy
6267 Delmar
St. Louis, Missouri 63130

Energy Times
P.O. Box 261
Dekalb, Illinois 60115

765 Hot Line
Newsletter of the Citizens for Safe Power Transmission (CSPT)
P.O. Box 351
Red Hook, New York 12571

It's About Times
Abalone Alliance Newsletter
944 Market Street   Room 307
San Francisco, Ca. 94102

Mid-Hudson Nuclear Opponents, Inc.
P.O. Box 666
New Paltz, New York 12561

Northern Sun News
Twin Cities Northern Sun Alliance
1519 E. Franklin
Minneapolis, Minn. 55404

Nuclear Times   A joint publication of
Crabshell Alliance and Live Without Trident
79 Yesler Way
Seattle, Washington 98104

Opihi Alliance for a Nuclear Free Hawaii
P.O. Box 26124
Honolulu, Hawaii 96825

Potomac Alliance
Natural Guard Fund
P.O. Box 9306
Washington, D.C. 20005

SEA Alliance News
Safe Energy Alternatives
324 Bloomfield Avenue
Montclair, N.J. 07042

The Crab Sheet
News of the Safe Energy Movement
Chesapeake Energy Alliance
609 Montpelier Street
Baltimore, Maryland 21218

Three Mile Island Alert
315 Peffer Street
Harrisburg, Pennsylvania 17102

Valley Environment
P.O. Box 2002
Evansville, Indiana 47714

NATIONAL GROUPS

Critical Mass Energy Project
P.O. Box 1538
Washington D.C. 20013
(Publishes *Critical Mass Journal* and provides information on
Congressional nuclear legislation and local, national and
international opposition to nuclear power)

Environmental Action
1346 Connecticut Ave., N.W.   Suite 731
Washington D.C. 20036
(202) 833-1845

Environmentalists for Full Employment
1536 Sixteenth St. N.W.
Washington D.C. 20036
(202) 347-5990
(Provides information on the relationship between nuclear
power and employment, solar and alternative energy and
employment. Publishes books and brochures.)

Friends of the Earth
124 Spear St.
San Francisco, Ca. 94105
(Publishes *Not Man Apart*, a monthly devoted to environmental issues and with an update on nuclear power.)

Media Access Project
1609 Connecticut Ave. N.W.
Washington D.C. 20009
(202) 232-4300
(Public interest law firm providing free legal counsel and
representation to citizen groups on Fairness Doctrine and
other media issues)

National PIRG
1129 21st St. N.W.
Washington D.C. 20036
(Has affiliated organizations in 25 states, most of which are
actively involved in safe energy activities. For information
on the Public Interest Research Group in your area, write
the national office.)

Natural Resources Defense Council
2345 Yale Street
Palo Alto, Ca. 94306
(Provides information on legal strategies for protecting
the environment.)

Nuclear Information and Resource Service (NIRS)
1536 16th St. N.W.
Washington D.C. 20036
(202) 483-0045   or toll free: (800) 424-2477
(Provides information on various aspects of nuclear power,
including "How to Pass A Transportation Ban". Offers
a toll free nuclear information service and publishes
Groundswell.)

NUCLEAR Transportation Project
American Friends Service Committee
P.O. Box 2344
High Point, N.C. 27261
(919) 483-0045
(Provides information on the transportation of radioactive
wastes, including a newsletter on the latest developments
in this phase of the fuel cycle.)

Residential Utility Consumer Action Group
(RUCAG)
P.O. Box 19312
Washington D.C. 20036
(Information on how rate payers can monitor the utilities)

Solar Lobby
1001 Connecticut Ave. N.W.   5th floor
Washington D.C. 20036
(Promotes national solar and conservation legislation.)

Task Force Against Nuclear Pollution
P.O.B. 1817
Washington D.C. 20013
(Franklin Gage, coordinator of the task force has been
working for several years on lobbying congress with the
clean energy petition.)

Union of Concerned Scientists
1025 15th
Washington D.C. 20005
(Independent technical studies relating to nuclear power
plant safety, strategic arms race, radioactive waste disposal
options and energy policy alternatives.)

WISE
1536 Sixteenth St. N.W.
Washington D.C. 20036
(Provides news from the international anti-nuclear/safe
energy movement.)

NATIONAL PERIODICALS

Critical Mass Journal
P.O. Box 1538
Washington D.C. 20013
(A monthly on safe and efficient energy and citizen action against nuclear power)

Energy
Energy Awareness Center, Inc.
P.O. Box 711
Woodstock, New York 12498
(A news digest of nuclear hazards versus alternative energy)

Groundswell
Nuclear Information and Resource Service
1536 16th St. N.W.
Washington D.C. 20036
(A resource journal for energy activists and information on how to oppose nuclear power)

New Age
32 Station St.
Brookline Village, MA. 02146
(Health issues, spiritual and educational concerns, energy issues and nuclear power updates)

New Roots
Box 548
Greenfield, MA. 01302
(Regional but growing. Devoted to health, science, politics and energy issues.)

Not Man Apart
Friends of the Earth
124 Spear St.
San Francisco, Ca. 94105
(Published biweekly. Devoted to environmental issues, with a "nuclear blowdown" section describing events, strategies, setbacks and progress in the anti-nuclear movement.)

People and Energy
Citizens, Energy Director
1413 K St. N.W. 8th floor
Washington D.C. 20005
(Excellent bimonthly news magazine)

Rain: A Journal of Appropriate Technology
2270 N.W. Irving
Portland, Oregon 97210
(Published monthly)

The Mobilizer
Mobilization for Survival
3601 Locust Walks
Philadelphia, Pa. 19105
(215) 386-4875
(Devoted to peace issues.)

The Power Line
Utility Project
Environmental Action Foundation
724 Dupont Circle Bldg.
Washington D.C. 20036
(Monthly newsletter on utility news and what citizens are doing to organize around utility issues)

The Progressive
408 W. Gorham St.
Madison, Wisc. 53703
(Peace, nuclear power and exposes)

RESOURCES ON NUCLEAR/ALTERNATE ENERGY

Anthropology Resource Center (ARC)
P.O. Box 90
Cambridge, MA 02138

Border Crossings
Box 146
Turners Falls, MA 01376
(Smiling Sun resources – "Nuclear Power—No Thanks" in 9 languages)

Citizens Energy Project (CEP)
1110-6th St. N.W. No. 300
Washington D.C. 20001
(Good literature on nuclear, solar and alternative energy)

Donnelly/Colt
Box 271
New Vernon, N.J. 07976
(Produces and distributes a variety of buttons and bumperstickers)

Environmental Action Reprint Service (EARS)
Box 545
La Veta, CO 81055
(A full catalog of materials on nukes and alternatives)

Feminist Resources on Energy and Ecology (FREE)
Box 6098
Teall Station
Syracuse, N.Y. 13217
(Literature, T-shirts, buttons)

Larry Fox
P.O. Box M
Valley Stream, N.Y. 11582
(Buttons, bumperstickers, balloons, etc.)

Mobilization for Survival (MFS)
3601 Locust Walk
Philadelphia, Pa. 19104

Nuclear Information and Resource Service (NIRS)
1536 Sixteenth St. N.W.
Washington D.C. 20036
(Literature and reports on nuclear energy and conservation)

Pacific Studies Center
867 W. Dana Street   No. 204
Mountain View, Ca. 94041

Progressive Foundation
315 W. Gorham
Madison, Wisc. 53703

Religious Task Force (MFS)
198 Broadway
New York, N.Y. 10038
(Assorted resources on nuclear power/weapons)

RESOURCES IN CANADA

The Birch Bark Alliance
OPIRG - Peterborough
c/o Trent University
Peterborough, Ontario
K9J 7B8
("Ontario's Voice of Nuclear Concern," a monthly newspaper)

Briarpatch
1409 10th Avenue
Regina, Saskatchewan
S4P OE2
(Saskatchewan's Independent monthly newsmagazine)

RESOURCES/UTILITIES

Utility Action Guide
Resource Materials on the Electric Power Industry
Claudia Comins
Environmental Action Foundation's Utility Clearinghouse
available from:
Environmental Action Foundation
724 Dupont Circle Building
Washington D.C. 20036
(202) 659-1130

RESOURCES FOR CHILDREN

Coloring Book
Women Concerned About Nuclear Power Plants
4108 Meadow Hill Lane
Fairfax, Virginia 22030

ALTERNATIVE RESOURCES

Alternative Energy Developments
The Rhino Publishing Company
P.O. Box 3214
Thousand Oaks, Ca. 91359

Citizens' Energy Project
1413 K Street N.W.   8th floor
Washington D.C. 20005

Consumer Action Now (C.A.N.)
355 Lexington Avenue
New York, N.Y. 10017

Food for Thought Books
325 Main Street
Amherst, MA 01002
(413) 253-5432

The National Center for Appropriate Technology (NCAT)
P.O. Box 3838
Butte, Montana 59701
(406) 494-4572

Northeast Appropriate Technology Network, Inc.
P.O. Box 548
Greenfield, MA 01301

Solar Concepts
Cornerstones Foundation
54 Cumberland Street
Brunswick, Maine 04011

Solar Energy Information Locator
Solar Energy Research Institute
1536 Cole Blvd.
Golden, Colo. 80401
(303) 231-1158 (Document Distribution Service)

Total Environmental Action, Inc.
24 Church Hill
Harrisville, N.H. 03450

"To X-Ray or Not to X-Ray?"
Whole Person Care
Mark Tager/Charles Jennings
Victoria House Publishers
2218 N.E. 8th
Portland, Oregon 97212

RESOURCES ON NUCLEAR POWER/WEAPONS

Coalition For A New Foreign and Military Policy
120 Maryland Ave. N.E.
Washington D.C. 20002

Institute for Defense and Disarmament Studies
251 Harvard St.
Brookline, MA 02146

Mobilization for Survival
3601 Locust Walk
Philadelphia, PA 19104

National Action/Research on the Military Industrial
Complex (NARMIC)
A Project of the American Friends Service Committee
1501 Cherry St.
Philadelphia, Pa. 19102
(215) 241-7175

Religious Task Force
Mobilization for Survival
198 Broadway
New York, N.Y. 10038

SANE
318 Massachusetts Ave. N.E.
Washington D.C. 20002
(202) 546-4868

War Resisters League
339 Lafayette Street
New York, N.Y. 10012

Women's International League for Peace and Freedom
1213 Race St.
Philadelphia, Pa. 19107
(215) 563-7110

Women Strike for Peace
5539 West Pico Blvd.
Los Angeles, Ca. 90019

HOW TO USE CIVIL DISOBEDIENCE TO OPPOSE NUCLEAR POWER

Civil Disobedience is a phrase that carries a negative charge for many people, conjur--ing visions of helmeted police, sirens, tear gas and limp bodies being hauled off to waiting vans. People ask: "Why do they do it?" The answer is that civil disobedience is nearly always the result of having tried, sometimes for decades, to work within a nonresponsive system. The people who marched on Selma with Martin Luther King, for example, had concluded that one hundred years of working within the system had not given black Americans their civil rights or eliminated the vestiges of twentieth century slavery. Surely, they said, in a nation where the descendants of slaves still rode in the back of the bus, went to segregated schools and lived in segregated neighborhoods, it was time to take some form of dramatic action in the name of freedom.

To those of us who believe that nuclear power is a threat to our planet and who have found through experience that the courts rule consistently in favor of utilities, civil dis--obedience is a moral imperative. We are told that building a nuclear plant which will emit low level radiation is not an act of violence because the utilities which finance these plants are operating within the law: Land has been purchased, construction and operating permits obtained and the blessing of various agencies secured. All this is done within the law. And so, once sanctified as private property, the monster is shrouded with all the pro-tection of a military installation. And if we express our concern for the environment by sitting, standing, walking, sleeping or in any way trespassing on this sancrosanct piece of ground, we are breaking the law—we are being violent.

But this Orwellian logic has not prevented opponents of nuclear power from occupying construction sites, scaling fences and lying on railroad tracks, acts for which they have been arrested, jailed, fined and sometimes beaten. Demonstrators have been denied a trial by jury, refused permission to use the competing harms defense,* and tried sepa-rately from friends or affinity group members. In the words of John Gofman, they have discovered that "Justice and the law are not synonymous."

In the long struggle ahead, civil disobedience will undoubtedly be used again and again to challenge the utilities' right to commit legalized murder, but it is important that we not romanticize this strategy. The essential question today is not how many times we are willing to go to jail for our convictions, but how our time, money and energies can best be used to build a broad based movement against nuclear power.

For people interested in direct nonviolent action, this section offers a broad overview of the history of nonviolence and its place in the anti-nuclear struggle, provides those who may have not yet been arrested or processed through the courts with an idea of what it is like at the trials of the fence jumpers and offers suggestions on how to prepare for one's day in court.

---

*Defense of Necessity/Choice of Evils exists when the harm resulting from compliance with the law would exceed the harm resulting from breaking the law, providing: (1) The harm is imminent, (2) alternatives are unavailable, and (3) the "illegal" action was reasonable. In the nuclear power area some courts have been reluctant to recognize the clear and imminent danger posed by plants and have refused to allow this defense in civil disobedience situations. At common law this defense only dealt with imminent dangers from obvious and generally recognized hazards. The Model Penal Code, which many states follow, provides that this defense is available, but only in the absence of an alternative legislative solution. For example, in many states the governor has the power to shut down a plant if it is determined to be an imminent danger.

from The Anti-Nuclear Legal Project's "pro se" Handbook.

NONVIOLENCE THEORY AND STRATEGY FOR THE ANTI-NUCLEAR MOVEMENT

By Gordon Faison and Bob Irwin

Movement For A New Society is a Philadelphia based organization that has been involved in the anti-nuclear movement since 1973. MNS publishes books and pamphlets on a variety of social issues such as health care, women's rights and nonviolent action. MNS also provides training to groups and individuals in nonviolent action.

In this article, MNS members Bob Irwin and Gordon Faison discuss the history of nonviolent action and how it can be used to oppose nuclear power. They also provide answers to some of the most commonly asked questions about nonviolent action.

The present state of the people's movement against nuclear power and nuclear weapons is exciting and promising. The spread of groups using nonviolent direct action has demonstrated how wide and deep is the opposition to deadly nuclear technology and the priorities and needs of the ruling power structure. The development of nonviolent direct action campaigns as a principal means of struggle has greatly expanded the ranks of those taking an active and committed part.

For new participants in these campaigns, nonviolence may often be mysterious, controversial, or both. Our aim is to provide a short introduction to nonviolence theory and strategy which can help dispel some of the mystery and clarify the controversies. Although in historical terms active nonviolence is still comparatively young, it has already proven to be a significant form of struggle. Now that the theory and dynamics are beginning to be understood, lessons have been learned which we should not ignore, for they can help us advance toward our goal of a better and more secure human future.

In the first part of this paper we survey the history, methods, and varieties of nonviolence. In the second part, we discuss its basic theory and dynamics and deal with some strategic issues. In the final section we suggest answers to some of the questions which may arise among participants in nonviolent struggle.

Within our severe limitations of space, we have said relatively little about nonviolent personal philosophies. We have emphasized nonviolence as a technique lest we otherwise seem to imply one must adopt such a philosophy before taking part in nonviolent struggle. Such involvement can raise important questions of motivation and values. We strongly encourage people to explore these further.

"The major advances in nonviolence have not come from people who have approached nonviolence as an end in itself, but from persons who were passionately striving to free themselves from social injustice." (Dave Dellinger, "The Future of Nonviolence.") The current renewal of interest in nonviolent tactics and strategies comes out of popular struggles. We

write as involved participants to increase the effectiveness of these struggles. We urge all who read this paper to take part in study training and nonviolent action and to consider carefully how we all can contribute toward shaping a more humane, more just society.

Although we have consulted with nonviolent activists around the country, this short paper is not meant as the last word on anything, and there is much we have had to omit.

HISTORY, METHODS, AND VARIETIES OF NONVIOLENCE

Nonviolent action is a means of social struggle which began to be developed in a conscious way only in the last several decades. It does not rely on the good will of the opponent but instead is designed to work in the face of determined opposition or violent repression. It is not limited to any race, nationality, social class, or gender and has been used successfully in widely varying political circumstances.

Nonviolent action is not simply any method of action which is not violent. Broadly speaking, it means taking action that goes beyond normal institutionalized political methods (voting, lobbying, letter writing, verbal expression) without injuring opponents. Nonviolent action, like war, is a means of waging conflict. It requires a willingness to take risks and bear suffering without retaliation. On the most fundamental level, it is a means by which people discover their social power.

Nonviolent action takes three main forms: protest and persuasion, noncooperation, and intervention.

The first category (protest and persuasion) includes such activities as speech-making, picketing, petitions, vigils, street theater, marches, rallies, and teach-ins. When practiced under conditions of governmental tolerance, these methods can be comparatively insignificant. When the views expressed are unpopular or controversial, or go against government policy, even the mildest of them may require great courage and can have a powerful impact.

The second category (noncooperation) involves active noncooperation. In the face of institutional injustice, people may refuse to act in ways which are considered "normal"--to work, buy, or obey. This largest category of nonviolent action includes refusal to pay taxes, withholding rent or utility payments, civil disobedience, draft resistance, fasting, and more than fifty different kinds of boycotts and strikes. Noncooperation can effectively halt the normal functioning of society, depending on the type of action employed and how widespread its use becomes.

Finally, there is the third category (intervention) which can be defined as the active insertion and disruptive presence of people in the usual processes of social institutions. This can include sit-ins, occupations, obstructions of "business as usual" in offices, the streets, or elsewhere, and creation of new social and economic institutions, including the establishment of parallel governments which compete with the old order for sovereignty. These methods tend both to pose a more direct and immediate challenge than the other methods described earlier and to bring either a quicker success or sharper repression.

These actions, taken from a list of nearly 200 methods compiled by researcher Gene Sharp, are plainly in the mainstream of the contemporary world. Virtually everyone has heard of these kinds of actions, and literally millions of people in the U.S. alone have taken part in one or more of them.

But what is the relation of these diverse actions to "nonviolence"? Most people involved in them do not believe in "nonviolence"—and what does it mean to "believe in nonviolence"? What is the difference between "pacifism" and "nonviolence"? In fact, there are several distinct types of principled nonviolence and failure to distinguish among them quickly leads to confusion.

The first sizable groups in the modern world who attempted to live their nonviolent ideals were small "non-resistant" Christian sects, such as the Mennonites and Anabaptists, who in times of war refused conscription into the army and bore punishments laid on them without resisting. Otherwise such groups were generally law-abiding, desiring to be left to pursue personal salvation. Where these groups still survive today, they rarely use the nonviolent methods mentioned above.

A second, more worldly nonviolence, which may be called "active reconciliation," is subscribed to by many Quakers and individual pacifists. They particularly aim to reconcile parties in conflict, to aid victims of war and poverty and to persuade by education and example rather than by coercion. Many programs of the Quaker-initiated American Friends Service Committee exemplify this viewpoint, such as its aid and self-help programs and promotion of dialogue on Middle East issues. Gene Sharp observes that "persons sharing the active reconciliation approach often prefer a rather quietist approach to social problems, disliking anything akin to *agitation* or *trouble*. Some of them may thus oppose nonviolent action (including strikes, boycotts, etc.) and even outspoken verbal statements, believing such methods to be violent in spirit. . ." Such conservative views are less prevalent among pacifists today than formerly. Many from

this tradition have gone on to make major contributions to nonviolent action.

A third category of adherents of nonviolence can be called advocates of "moral resistance." Although advocating and engaging in education and projects promoting human cooperation, they frequently lack an overall social analysis or comprehensive program of social change. Nineteenth century Americans agitating for the abolition of slavery were among the first to articulate "moral resistance." Many activities of the civil rights and anti-Vietnam War movements, such as sit-ins, marches, draft refusal, blockage of ammunition shipments, and obstruction at induction centers, reflected this outlook, shared by many individual pacifists.

These three varieties of nonviolence suffer from significant limitations. There has been considerable growth in the methods that we now call nonviolent. These means of struggle were invented in the context of some of the major conflicts (as in the American colonies) and struggles between labor and capital. The notion of civil disobedience and the value of nonviolent resistance were spread by writers like Thoreau and Tolstoy. But pacifists had abolished neither war nor injustice. They lacked a sufficiently powerful method of actively pursuing their goals, one that could harness human courage, energy, idealism, and solidarity.

The career of Mohandas Gandhi (1869-1948) marked a watershed in the development of nonviolent struggle. In leading the struggle for Indian independence, Gandhi was the first to conduct a series of campaigns toward long-term goals. Deeply religious, practical, and experimental in temperament, Gandhi was a shrewd, tireless, and efficient organizer who united cheerfulness with unshakable determination. He was not only a political strategist but a social visionary. His approach to social change had three main elements: 1) self-improvement—the effort to make oneself a better person, 2) a constructive program—concrete work to create the new social order aimed at, and 3) campaigns of resistance against evils that blocked the way forward, such as the caste system and British colonial exploitation. Gandhi's success in linking mass action with nonviolent discipline showed the enormous social power this form of struggle could generate. While his contribution was overwhelmingly positive, it is also true that his experimental, unsystematic approach and personal charisma make it difficult to disentangle those aspects of his approach peculiar to the requirements of Indian society, or which expressed his personal eccentricities, from those aspects of nonviolent action of possible universal application.

It is through nonviolent direct action campaigns in the tradition of Gandhi that most people in the U.S. have become aware of nonviolence and nonviolent methods. In fact, despite the many violent aspects of American history of which we have become increasingly aware in recent years, the U.S. has its own native tradition of nonviolence. Staughton Lynd has noted that "America has more often been the teacher than the student of the nonviolent ideal." (*Nonviolence in America*)

Nonviolent currents in American history can be divided into three parts:

1) *The use of methods which in retrospect we recognize as nonviolent.* The movement for women's rights during the nineteenth century used civil disobedience, tax refusal, and public demonstrations. Alice Paul's Woman's Party used the vigil and hunger strike to exert pressure on behalf of women's right to vote. During the Great Depression of the 1930's the sit-down strike was used as a way to force recognition of trade unions. Less well-known, but highly significant, was the plan of struggle called the Continental Association, adopted in October, 1774. Delegates from the thirteen colonies agreed on a program which included both economic boycotts and social boycotts and other sanctions against those reluctant to comply. Their program was the major pre-Gandhian campaign to include planned strategic phasing of the struggle.

2) *The participation of adherents of nonviolence in important struggles.* Examples already mentioned include the struggle for the abolition of slavery, for women's suffrage, for the rights of labor, and for civil liberties. Many organizations and institutions grew out of pacifist commitments, including Brookwood Labor College (the first residential labor college in America), National Conference of Christians and Jews, American Civil Liberties Union, American Committee on Africa, Society for Social Responsibility in Science, and the Congress of Racial Equality. Many fought for racial justice, others for admission of Jewish refugees during the 1930's. Opposition to war and violence logically drew people to work actively against other kinds of injustice. Although frequently undramatic, the work accomplished by such people has contributed substantially to the betterment of the society.

3) *Actions and campaigns undertaken or directed by explicitly nonviolent leadership.* During World War II and shortly thereafter, militant pacifists succeeded in ending racial segregation in prisons where they themselves were held, and took part in the first "Freedom Rides" to desegregate interstate transportation. The most dramatic nonviolent actions of the 1950s were several voyages into nuclear testing areas by small vessels with pacifist crews. In a time when nuclear war seemed a fate humanity was powerless to overcome, these actions gave expression to the widespread yearning to act in some way against the madness of testing and the arms race. Although in each case the boats were prevented from reaching their destinations, the powerful symbolism of the voyages succeeded in boosting the morale of the anti-nuclear movement, thus giving a real impetus to the public sentiment which resulted in the 1963 test-ban treaty.

Equally important was the inspiration provided by nonviolent activists through their examples of courage and the taking on of personal responsibility for institutional injustice. Historians of the New Left have noted that it consciously adopted issues, tactics and moral postures from the nonviolent tactics of personal witness and mass civil disobedience. But it was the movement of black people for civil rights and an end to racial oppression which imprinted the idea of nonviolence on the American consciousness. The bus boycott in Montgomery, Alabama which began in December, 1955, when Rosa Parks refused to surrender her seat to a white passenger, grew to include an alternative transportation system and ended with the desegregation of the entire bus system. An eloquent young minister, Dr. Martin Luther King, Jr., attained national prominence as a spokesperson in the struggle, demonstrating that nonviolence not only could win significant victories in India but also in the U.S. despite racial violence and intimidation.

THE PRACTICE OF NONVIOLENCE

In 1960 a new wave of activity began when the first sit-in was undertaken by four black college students in Greensboro, North Carolina (one of whom had just been reading a comic book about the Montgomery campaign issued by the pacifist Fellowship of Reconciliation), who decided to fight the refusal of service at a local lunch-counter. The action spread rapidly and spurred a wave of related actions in other places of public accommodation. Under the pressure of actions by many small groups of activists whose demands were widely perceived as just, new court decisions began to legitimize the changes for which people were struggling. As campaigns continued in many places, loosely coordinated by such groups as the Southern Christian Leadership Conference (SCLS) and the Student Nonviolent Coordinating Committee (SNCC), resources would be shifted at times of crisis to certain cities that became focal points, such as Birmingham in 1963 and Selma, Alabama, in 1965.

King's important role as a spokesperson and moral symbol of the struggle has frequently led to an underemphasis of the grass roots, decentralized nature of the movement, whose heart was the decision by thousands of people to risk their security and often their lives on behalf of the cause and to grow toward a greater fulfillment of their own potential in pursuit of justice and human community.

The civil rights movement had enormous and lasting impact. It affected both blacks and whites through the legal and institutional changes it brought, and it also created a body of people with a shared moral and political background from which they could move on to challenge other injustices like the Vietnam War, imperialism, poverty and sexism. This achievement was often minimized by those who became increasingly radicalized by their experiences when they saw clearly how much more remained to be done—that they were engaged in more than correcting a flaw in an otherwise healthy system. Those entering the movement for social change later, sometimes, took for granted the gains which had been made at such cost. The death of Dr. King in 1968 during the Poor People's Campaign, which had aimed to unite poor people of all races around economic issues, was a critical blow to a movement beset by other problems as it attempted to move forward. Although the civil rights movement and Dr. King were moving into wider arenas, the experience can still serve as a reminder of the limitations of a nonviolent movement focusing on a single issue, be it war or racism, rather than aiming at the revolutionary transformation of the whole society.

NONVIOLENCE THEORY AND STRATEGY

The Spreading of Anti-Nuclear Campaigns

Before discussing the theory of nonviolence and various strategic issues, it is useful to review how the adoption of nonviolent direct action as a method of struggle by American anti-nuclear groups occurred. The underlying cause was simply the general situation. The structural conditions for resorting to nonviolent action were present: more conventional political and legal channels appeared blocked, yet people were unwilling to abandon their goals. More particularly, Clamshell Alliance organizers, buoyed by earlier struggles in their region and determined to persevere, found a model that seemed relevant in the experience of nuclear opponents elsewhere (Wyhl, West Germany): a nonviolent occupation.

Nonviolent action frequently spreads this way. People hear of or remember an action that seems relevant and imitate or adapt it to their own circumstances. If it is successful (or meaningful, regardless of immediate success), it may spread further, as did the sitdown strikes of the 1930's, the sit-ins of the 1960's and the housewives' meat boycott in the early 1970's.

This process need not be spontaneous; it can be deliberately fostered. In a 1972 speech entitled "De-developing the U.S. Through Nonviolence," Movement for a New Society co-founder William Moyer proposed a strategy for a nationwide and trans-national movement against nuclear power. Rather than starting by forming a national coalition of sponsoring groups (a process with several disadvantages detailed in the article), "the campaign movement approach encourages groups to organize whatever local socio-dramas they believe to be creative and important. Small groups begin small projects in different places, joining others only when interests coincide. The key here is not the size of initial numbers, but the ability to organize a local campaign with drama, crises, and other socio-drama elements. Even when all these ingredients are present, however, there is no guarantee that a project will take off into a full-fledged movement.

"The strategy of the campaign-movement approach to nationwide efforts is that if enough independent socio-drama projects are begun, there will soon be one which reaches a take-off point, with much drama, crisis, publicity, and interest. At this 'crunch' point, some people in other projects and regions can temporarily join the socio-drama taking off, for experience and training, then return to their own area to start a similar campaign. People in the original socio-drama can also travel to other areas to help start new campaigns."

The strength of this approach has been spectacularly demonstrated by the Clamshell Alliance and the movement against the Trident submarine.

The Dynamics of Nonviolent Action

The conventional view of power is that some people have it and others don't. Power can reside in soldiers, authority, ownership of wealth, and institutions. The nonviolent theory of power is essentially different; rather than seeing power as something possessed, it argues that power is a dynamic social relation. Power depends on continuing obedience. When people refuse to obey rulers, the rulers' power begins to crumble. This basic truth is in a sense obvious, yet it took the dramatic historical episodes of Gandhi's civil disobedience campaigns to begin to establish a new model of power. In routine social life this truth is obscured, but events like Seabrook cannot be understood without it.

From the standpoint of the conventional theory of power, what happened at Seabrook in April/May 1977? Protesters came and occupied a dusty parking lot; they were arrested and taken away. Action and reaction. Beginning and end. Defeat—they were all in jail. In reality the picture was far more complex. Instead of two social actors at work—the Clamshell Alliance and Governor Thomson of New Hampshire—a whole range of intermediary forces was involved. Nonviolent action has significant effects on these forces, in ways we will describe, using Seabrook as our model.

Gene Sharp's later chapters in *The Politics of Nonviolent Action* offer an outline of topics we follow here: laying the groundwork for nonviolent action; challenge brings repression; solidarity and discipline to fight repression; and ways success may be achieved.

Laying the groundwork is fundamentally important. This means defining goals, choosing strategy and tactics, making contingency plans, training, etc. Nonviolence is not magic; it is a way of mobilizing the strength we have for maximum effectiveness. "Action and reaction" which seem to be the whole story are only the beginning. Along with the leading actors who clash with each other, there are also Clamshell members who are not committing civil disobedience but playing active support roles; potential participants who didn't feel enough urgency or sense of being needed to take part in that particular action; people who would like to see an end to nuclear power but don't plan to do anything about it; people oblivious to the issue; people hostile to "environmentalists who delay needed progress"; people who say lawbreakers should be punished, but will limit themselves to griping; on up to utility executives, the governor's staff, bank presidents, etc. There are also police and National Guardspeople whose job it is to counter the demonstrators, but whose personal attitudes may lie anywhere on the spectrum.

The actions of the main social actors potentially affect all these people. The outbreak of conflict draws attention to the issue. In an important respect the two sides are not fighting each other directly, but competing with each other for the support of the public. On an issue like nuclear power, where there are many reasons to be against it and where opposition is growing, almost any public discussion favors anti-nuclear forces. When nonviolent direct action has been initiated, pro-nuclear forces often try to portray the activists as a small, irresponsible group which is flouting the public interest. (This image in fact fits the *pro-nuke* forces all too readily!) A next step can be to invoke general principles like obedience to law to stir fears of disorder. But well-disciplined nonviolent action works against this diversion. Openly friendly and orderly activists pose little apparent threat to the public; with the onset of repression the question tends to arise, 'Why did they put themselves in such personal jeopardy? Is the repression justified?' If their behavior can be made to serve their cause in this way, the activists may be on the road to success. Attempts to distort the image of the activists may fail and backfire; e.g., when one of the best organized nonviolent actions in history is described as the work of "terrorists," even the unsympathetic are likely to raise a skeptical eyebrow.

To gain their desired result, agents of repression must make the activists lose their solidarity and abandon their goals. If they maintain solidarity and discipline, as they did at Seabrook, repression becomes ineffective. But solidarity alone does not bring success. That may come through a kind of "political ju-jitsu," in which the repressive efforts themselves tend to shift the balance of power toward the nonviolent activists. People on the nonviolent side increase their level of involvement, while those allied with the opponent may reduce their support or switch sides. Shifts of attitude are important as well as shifts of behavior, because both sides adjust their action according to how they gauge their support. Costly incarceration posed a question to New Hampshire residents-- "Is it worth it?" --a question to which they were bound to answer "no" before long.

Nonviolent action is not dependent on the opponent's being repressive or making mistakes. It is not stymied when the opponent is moderate and conciliatory. Most of the methods mobilize political strength regardless of the opponent's response.

This brings us to the question of how nonviolent action may attain its goals. Three main ways have been identified: conversion, accommodation, and nonviolent coercion. Conversion means that the opponent has a change of heart or mind and comes to agree with and work toward the activists' goal. At the top of the social structure, this is fairly unlikely, but significant instances may occur: for example, Daniel Ellsberg, who released the Pentagon Papers after being converted to opposition to the Vietnam War; Bob Aldridge, who left his job as chief missile designer for the Trident submarine in order to speak out against the growing threat of nuclear catastrophe. At the other extreme is nonviolent coercion, where the activists have it directly in their power to frustrate the opponent's will. One example is the refusal by all workers to work on construction projects which a union has declared unecological (Australia's "green bans"), another was the invention of the "search and avoid" missions by GIs in Vietnam who did not want to risk their lives in an unpopular war. Most commonly the outcome is determined by an intermediate process. Accommodation

means the opponents give in, partly or completely, not because they have changed their minds, and not because they are completely powerless, but because continuing the struggle at that point would probably mean further erosion of support. Concessions may also be granted to halt the consciousness-raising process of struggle which would lead people to discover how much power they really have.

QUESTIONS AND ANSWERS CONCERNING NONVIOLENT ACTION

Q. Nonviolent discipline is oppressive, and who are we to tell other people they can't join us if they don't agree with the discipline?

A. To be effective, any approach to social change has requirements. Driving a car places restraints. A driver may not drive from the front bumper or trunk if wishing to proceed, nor may one drive on the wrong side of the road without serious consequences.

Gripes about nonviolent discipline frequently mask a more serious question: "Nonviolence really doesn't work, so why bother with all the training and discipline anyway?" Nonviolent discipline is rooted not only in a theoretical perspective but also in the knowledge that nonviolence *works*. It may or may not work in a given instance (and the adequacy of the strategy chosen is crucial), but it cannot be tested unless the participants in nonviolent campaigns are willing to abide by and use nonviolent tools in a disciplined, organized, and empowering fashion. Given the increasing size of anti-nuclear demonstrations and the increasing power of the anti-nuclear movement, the unpredictability of the government's responses, the presence of *agents provocateurs*, and the consequences of activist-instigated violence for a long-term campaign, people who wish to participate should at the very least be willing to suspend judgement as to the efficacy of nonviolence and abide by a commonly shared code of discipline. Otherwise they should be firmly requested not to participate, so that the large majority who wish to experiment with the use of militant nonviolence are free to do so.

Q. Why do we need to inform our opponents of what we plan to do?

A. Being open about one's intentions may seem odd when engaged in serious struggle. Deception may seem to offer advantages. Nevertheless, openness is crucial for nonviolent action.

The negative effects of secrecy are all too obvious. Secrecy results in inefficiency, authoritarianism, and mistrust simply because of the need to cover up much of what is planned from our allies. Dependence on secrecy opens a movement to disruption by planted provocateurs and informers. Secrecy thus contributes to fears of betrayal; moves toward secrecy often come when a movement is losing self-confidence and weaken it further, reducing its numbers and attracting people of a furtive, conspiratorial disposition.

Equally important are the positive effects of openness. It is consistent with our purpose of educating the public about issues, and with the kind of society we hope to build. Openness creates a positive image in the public mind. It says we believe our actions are legitimate and that we expect others to think so too, thereby encouraging them to take this view. Openness increases the morale and self-respect of participants: our open and participatory democracy contrasts with the secrecy and high-handedness of our opponents.

One aspect of openness deserves particular attention: relations with police and other authorities. It has been asserted that police are not impartial enforcers of justice but rather agents of an unjust system whose authority should therefore not be respected. Working with police by informing them of our plans is interpreted as making their job easier, accepting their authority, and thus lending support to the system we should be fighting. The first premise is accurate here, but not the conclusions. Although agents of a system may sometimes symbolize and seem to embody it, they must not be confused with the system itself or the real power structure. In any case, there is no reason for hostility towards them on principle, any more than towards others whose social activity is contrary to their own best interests. The desire of those unsure of their own radicalism to seek reassurance in hostile "militant" attitudes cannot be accepted as a guide, for the truly revolutionary slogan is not "Off the pig!" but "Join us!"

Q. Nonviolence is okay for men, but women need to take power.

A. Barbara Deming has asserted that nonviolence is an essentially androgynous method of waging conflict. Nonviolent action incorporates values traditionally viewed as "masculine" in that it assumes the importance of active involvement in conflict and of involving oneself in social situations and wielding power responsibly. It incorporates traditionally "feminine" values in that it actively promotes a continuing concern for individual welfare, social healing and respect for life.

For men, nonviolence provides a context within which they can carefully assess whether their actions and thinking are oppressive to women, and a framework in which men can learn to be fully nuturing human beings. For women, nonviolence legitimizes and encourages a very real need to be assertive in social conflicts, without giving up the value of nurturing and an insistence on responding humanely to all human situations.

Given that both men and women are strongly conditioned by oppressive institutions to act the way they do, nonviolent action provides a model for challenging old behavior patterns which relegate us to being less than the people we really are capable of being.

Q. What about property destruction? Can it be nonviolent?

A. The risk in property destruction is that it moves toward the logic of violence. If we are determined to destroy some piece of property, will we be willing to injure some person who stands in our way? The dangers of property destruction are substantial. It may provide a readier pretext for repression. It can be a way of slipping toward violence, reflecting a loss of confidence in one's chosen means and an inclination to waffle between two contrary strategic choices. Such ambiguity can encourage violence by other participants and prove fatal to success.

Property destruction can, in certain circumstances, be an effective tactic but must always be evaluated according to whether it will be understood primarily as a "challenge in *human terms* by human beings to other human beings." Effective use of property destruction is therefore only likely where haphazard and undisciplined destruction is avoided and any destruction is completely open and subject to careful and deliberate control.

Q. But we've tried nonviolence and it didn't work.

A. "We've tried nonviolence" often translates into "I'm frustrated and angry, and violence is quicker anyway." This is an understandable position, given that we are raised in a "ready-mix" culture which encourages us to resort to violence when we are feeling desperate.

It is important to separate our feelings of desperation from our best thinking. Compelling and unrealistic hopes for a quick "victory" impede the development of any kind of effective strategy.

Nonviolent struggle does not guarantee success any more than violent struggle does. It is crucial to apply similar criteria when evaluating the effectiveness of these struggles, as is not usually done. Failures of violent struggle are usually attributed to poor strategy, insufficient materials, and bad morale. In contrast, the failure of a nonviolent struggle is usually attributed to *nonviolence,* and *not to the way the struggle was conducted.* Similarly, the value and importance of nonviolent successes are minimized, while violent successes are exaggerated. Given that nonviolence is in what Dellinger calls the "Edison and Marconi" stage of development, we are impressed by the frequency of "success" and are excited by the possibilities of replacing essentially *ad hoc* tactics with more systematic and conscious militant nonviolent strategies.

A NONVIOLENT ACTION MANUAL

By William Moyer

William Moyer, one of the founders of the Movement for a New Society, lives in Philadelphia Life Center, a nonviolent training-action center for social change.

Nonviolent action is a powerful people's weapon for social change. Most people know nonviolence as a vague and mysterious method used by a few exceptional people in exceptional places, such as Gandhi in colonial India and King in racist Southern United States. Few people, however, realize that nonviolence has been used extensively and with great success throughout history by ordinary people in ordinary settings—and that this method is available to them today. It is available also to you, the reader. You can use it in your neighborhood, in your town, on your issues.

The purpose of this small manual is to help people use nonviolent direct action demonstrations and campaigns in their efforts to take charge of their own lives. It is brief and in outline style for easy use.

NONVIOLENT SOCIAL ACTION

What: Nonviolence is a particular kind of citizens action for social change. It is a method for social liberation: "Gandhi recognized that Governments and institutions often fall into decay, whence they become oppressive and tyrannous, and also that evil customs are capable of destroying even the power to recognize their evil nature. When such situations arise, nonviolent resistance and revolt may be absolutely necessary in order to awaken the social conscience." (Wellock, *Gandhi as Social Revolutionary,* TAFSC, p. 8)

Goals: To achieve social justice and equality; not to achieve personal gain. To achieve the maximum meeting of human needs (psychological, spiritual and physical needs). It considers all people of the world and future generations.

Special Aspects of Nonviolent Method: Openness; truthfulness; self-suffering and personal risk rather than hurting or endangering opponents; means are considered the ends in the making; it is truth-seeking (it doesn't claim to know the whole truth, just "relative" truth); it defines truth as the maximum meeting of human needs.

Who: Anyone seeking justice through love and action. It can be used by all who want to work for a better world, privileged and oppressed, who want to try living a whole liberated life now. People who see themselves as personal, social and political beings. People who accept nonviolent goals and methods.

Nonviolent Movements, Campaigns and Demonstrations: A *demonstration* is a particular event or action on a given day, such as a picket or rally. A *campaign* is a prolonged citizens' action made up of a series of demonstrations over a period of weeks and months. Some campaigns last six months and longer. A *movement* is a wide-spread citizens' action which can last many months or even many years. It is made up of many demonstrations and many campaigns, often in different places simultaneously. It gets many people moving on an issue. Some past and present movements in the U.S. are: voting rights, public accommodations, school integration, anti-Vietnam War, and anti-nuclear power plants.

THREE THEORIES OF ACTION

In order to think about or do any activity, it is helpful to have some ideas and guidelines on which to begin. In this section we will discuss three different theories which will be helpful to you in thinking about doing campaigns.

Myths and Secrets
(Based on Max Weber's Theory of Dominancy)

In every society, social systems distribute most of the benefits and resources to a relatively few people (the "positively privileged"). That the system--not individual effort--distributes resources and is responsible for gross inequalities has to be kept a secret, to keep the negatively privileged from rebelling and to keep the positively privileged from rejecting their advantages. The ways the system maldistributes benefits, therefore, Weber called *societal secrets.* To hide these secrets, societies develop a public set of beliefs, values, and ideologies called *societal myths,* which are exactly the opposite of societal secrets. The population is socialized to believe so strongly in the rightness of the myths that when societal secrets are revealed, they will take vigorous action to change the unjust condition. (Particularly if given specific opportunities to do so.) A key to change, therefore, is to reveal clearly societal secrets and their corresponding myths-- and give people specific things to do to correct them. This is similar to Gandhi's notion of "truth force": when people see the truth of an unjust situation, they will try to change it.*

Gandhi's Satyagraha (Truth Force) Satyagraha has three aspects: truth, nonviolent action & self-suffering.

Truth
Truth is the maximum meeting of people's needs. We only know "relative" truth; that is, truth as we see it from our limited perspective. There is power in truth.

Nonviolent Action
The *act* of holding on to truth in public conflict. Action which challenges conditions in a society where human needs are not being adequately met. Nonviolent action is action in which we test our relative truth. It is action in which we make our concerns for human needs into public issues. The issue must be important to human needs. . .
The *Process* is important. We must give our opponents every opportunity to convince us they are right. Listen, dialogue with everyone, especially opponents. Try to convince the opponent, not defeat him. We cannot hate our opponents. Our ends must be embodied in our means. We must be open in our activities—we must have no secrets.

Self-Suffering
We accept suffering. We must not harm our opponents or the police. We must help our opponents save face. Suffering is a substitute for violence; it reduces physical and psychological injury. Suffering helps keep the focus on our issue and makes us more human in the eyes of both opponents and public. We should not be "martyrs."

NONVIOLENT DIRECT ACTION SCENARIO (SOCIODRAMA)

Many action ideas are generated in strategy meetings through brainstorming, scenario writing by small buzz groups or other

---

*NOTE: Some examples of societal myths and secrets: *Voting rights movement:* A basic *value* in the U.S. is the right to vote. In the South it was kept a *secret* that the system did not let blacks vote. *Myths* such as blacks didn't want to vote were perpetuated. When these myths and secrets were exposed to the American public and roles to act were provided, there was an outcry for change. *U.S. Support of the Pakistani army and war against Bangleadesh:* A basic *value* in the U.S. is not to participate in massive genocide and overthrow democratic governments. The government's support of Pakistan was kept a *secret*. The *myth* was that the U.S. stopped all shipments. When the ships going to Pakistan were continuously exposed on TV, the public and long-shoremen complained and the government ended its shipments. *Economic distribution;* Equal Economic opportunity is a *value* in the U.S. That the private enterprise system and governmental action grossly maldistributes wealth and income-not individual work efforts-is a *secret.* The *myth* is that hard work is the source of income and wealth. This can only be changed when the American people see how the system & government does this maldistribution.

techniques. How do we know which ones to choose? How can we improve on them? How evaluate them? In this section we have some guidelines for developing and evaluating action ideas. We call it a "sociodrama" because our goal is to develop and enact a real-life drama based on the issues we are concerned about and the analysis we made earlier. The following are guidelines for devising a nonviolent direct action scenario or sociodrama:

Reduce our analysis to a picture in which we can enact our real-life situation
A sociodrama demonstration should reduce the problem to a picture; ideally, just a few words would be needed. In a real-life dramatization of the problem try to portray the following aspects of our analysis:
>The violation of a widely held belief
>Our long-range and short-range goals
>Reveal the societal secrets
>Reveal the societal myths accompanying the secrets
>Show our alternative (s).

This differs from street theater, which is a play in which actors impersonate the real situation and the participants. But there we are trying to act out the real life situation, providing roles for ourselves, opponents, public, police, media, government. In trying to devise a sociodrama scenario which accomplishes these goals, it is helpful to ask, "What actions would a full-person do in this situation?"

Example; Voting
In America everyone is guaranteed the right to vote. This simple fact was the basic reason for the voting rights movement in the South. It showed the violation of the widely held belief that everyone *could* vote. And it revealed the secret that blacks weren't allowed to vote; revealed the idea that blacks did not want to vote as nothing but a myth, and showed the long-range goal of civil rights for nonwhites and the short-range goal of blacks waiting to register to vote. Almost daily national television and newspapers showed blacks lined up to vote-and being stopped and often attacked by police. When the movement's message is in a picture form the effectiveness of media misinterpretations and distorted statements by the opposition is reduced.

Select immediate targets
Be clear about your targets and keep a focus on them. For example, the registration offices, Pakistani freighters, trains which carry bombs, nuclear plants and waste storage sites.

Choose a battleground
A specific geographical area should be chosen where much of the direct action occurs. For example, sidewalks in front of a courthouse, railroad tracks, construction sites, etc. This does not mean that some actions couldn't occur at other places or there couldn't be shifts in the battleground at strategic times or for variety. Especially at the beginning of a campaign it is often good to do actions at different places to test which places bring the best results.

Make actions dramatic with crises

Drama increases public interest and media coverage, which in turn increases the degree to which we reach the public with our message and are able to raise people's consciousness. Crises also help force people into thinking and acting; they also enable organizations and government bodies to make decisions faster. A sociodrama is the enactment of a moral and ethical violation, so a sociodrama presents a moral crisis to the community. In this way a sociodrama is like a "moral equivalent of war." One element that adds to a demonstration's drama is civil disobedience, people doing things which seem right or upholding a moral value, but which also places them in a position for arrest by police. Actions which involve civil disobedience and arrests (even if only a few arrests) often build up more interests and drama as they are repeated over a period of weeks. Drama and crisis is also built up by marching to the place of confrontation from an assembly place (best from a meeting in a church) some distance away. Creative, unusual, interesting, sacrificial and "dangerous" actions can also be interesting and dramatic. For example the small boats in front of giant freighters, or marching to the courthouse to vote or sitting on railroad tracks.

Repeat actions (and crises) over a period of time.

It takes time for a campaign to build up. It takes time for people and the press to hear about a campaign so they can join it or cover it; once a dramatic confrontation occurs the public and the press are ready for the next time. Each time a simple confrontation occurs it becomes more newsworthy up to a point. News stories snowball. That is, it takes a while for a situation or an issue to be considered "news", then the media will give more coverage. Therefore, it is best to have a series of actions over a period of time. Be wary of actions everyday or around the clock because they deplete energies too much and too fast. A good simple action can be repeated many times. For example, sitting at a restaurant, reading war-dead names at the capitol, blocking ships with canoes.

Give Everyone A Role To Play

Try to have something for everyone to do, including concerned public, press, police, hostile opponents, people at home (write or call their congresspeople), church groups, consumers, public officials, community groups, artists, children, old folks, etc.

Repeatable elsewhere

Actions should be devised so they can be done in other places by other people. Good actions often start in one place and automatically spread as people elsewhere hear about the action and decide to do it themselves. This is how the sit-ins, reading the names of war dead, Pakistani ship blockades, train blockades and many others became nation-wide movements.

Choose important, timely and repeatable issues which have a "handle"

The Pakistani freighter blockades, for example, were chosen because the war was going on and there was potential for famine which could induce calamities. The blockades were timely because the war was occuring that year only. Also, the freighters sailing into American ports provided a good handle to expose the secret that the U.S. was supporting Pakistan. If we just picketed in front of a government building it wouldn't have been very effective. Also, the blockades were repeatable in other ports.

Start Small

Good nonviolent actions can, and perhaps should, start small with 5, 10, or 20 people. Often there aren't many people to participate at first. It's good to develop experience so you can control a demonstration which has large numbers. Most successful campaigns started with very small numbers and grew. The sit-ins started with four students at a lunch counter. Even after a particular campaign has some large demonstrations because of a particular crisis or special event, it often returns to a low number of people, the regulars who carry it on. The point is that the numbers of people on demonstrations is cyclical—they go up and down.

DESCRIPTION OF TASKS

Every nonviolent direct action requires the execution of a number of tasks. At least one person should be responsible for each task, hopefully a committee would do each if there are enough people. If more than one person does a task, one specific person should be designated as "coordinator" of that task group and should be responsible for being sure the group meets, does the tasks, & keeps in touch with the campaign coordinators. At organization meetings it's a good idea to have all the tasks on a large flip chart with the names of the volunteers written next to each task and the task coordinator marked with an "X". People can volunteer for more than one task at a time.

Recruitment

EVERYONE should recruit. But one person or task group should be designated as recruitment coordinator. Make sure all potential people who might join the campaign are contacted for each demonstration. Develop a list of people who might demonstrate. Have a phone tree to let people know about upcoming events. (See How To Develop A Phone Tree) Keep a list of everyone who joins all demonstrations. (See Civil Disobedience section: Legal Coordinator's Checklists) In recruiting you might ask people to join a task group. Often people are willing to become involved if they are able to contribute more.

Negotiation team

The minimum number on the team should be 3 and between 5 and 7 is even better. At least several members of the team should be extremely familiar with the issues and the campaign. The group should be democratically chosen at a regular meeting of the entire group. It is good to have some rotation of members, but there should also be some continuity of membership--especially the key contact people with the opponent. The team should hold a planning meeting before every negotiation meeting to develop strategy and tactics for the upcoming meeting. No new policy decisions should be made by the negotiation team in its meetings with the opponents. Be honest, open, friendly, but firm towards the opponent. Be good listeners, learn the opponent's position and rationale; explain clearly the movement's position; have a dialogue. Ask challenging

questions. See what common ground there is or might be. Try to develop resolutions which don't sacrifice principles. If negotiations last a long time, send out a messenger to let demonstrators outside know what's happening. A recorder should write down key points for reporting later. Give a copy to the campaign "scribe". At the end of the negotiating meeting, tell demonstrators outside what happened.

SOCIODRAMA--Strategy and tactics

The strategy and sociodrama must be democratically developed and agreed upon by the entire group. But a specific strategy team is needed to meet separately (open to all who want to attend) to brainstorm strategy, do research about the opponent, the layout of the demonstration area, get a nearby church or meeting place to hold a meeting before and after the demonstration, if desirable and possible. Also, there should be alternative strategies and tactics if there are potential forces which would require changes in the actions, such as bad weather, police actions, etc.

Legal

Ascertain the legal aspects of the demonstration. What is legal and not legal? Get a lawyer for the event. The lawyer doesn't have to be at the demonstration, but should be on call. Hopefully legal services will be donated. Try to get a "movement" lawyer or one that serves the group rather than "representing" the group. Get legal opinions from the lawyer ahead of time about the legality of particular activities and their probable consequences. If arrests or heavy conflict might occur, have legal *observers*--that is, people who just observe the demonstration for purposes of being witnesses. They should be respected, honest, clearheaded people; they should write down key events as they happen or shortly thereafter; keep a record of times things happen, places where events occur, and who is involved in the action.

Leaflets

Write, design, print leaflets. It is better to have too many than to run out. Make sure the leaflets get to the demonstration, preferably before it begins, and are distributed by members of your organization who can answer questions about the leaflet's contents. Writing leaflets should be a group process, i.e. the content of the leaflet should be checked by the group before distribution. Leaflets should give what, when, who, why, where. Use language people can understand and accept--no radical jargon or dirty words. Be honest, clear, factual and brief. Include the name, address and phone number of your group somewhere on the leaflet. Avoid distributing fuzzy, smeared, poorly xeroxed leaflets. State the problem in boldfaced headlines, and in lower case give some evidence supporting the headline. State what people can do about the problem. Be concise. Too many words tend to annoy or intimidate people.

Signs

Make picket signs, and banners if desirable, for every demonstration. Have the signs and banners at the demonstration site *before* the demonstration begins. Slogans should be agreed to by the entire group or by a committee designated by the group. Signs should have large, black print on a white background as this gives the clearest and most readable picture. Many signs can be made quickly by cutting out letters on one sign and using this sign as a stencil. Ink should be waterproof. One to five words make the best slogan for the largest lettering.

Logistics

Be sure you are in agreement about a meeting place and recheck this decision before the demonstration date. Make a thorough study of your transporation needs, how many cars will be required to carry your group's members to and from the demonstration. If you do not plan on getting arrested at the demonstration, agree to meet at a location nearby. If at all possible check the demonstration site ahead of time to determine whether there will be toilet facilities, water, food, day care for children and a medical team in case of special problems or injuries. Coordinate your efforts with other grass roots groups. For example, two or more groups could agree to provide food and water for a certain number of demonstrators. The attention paid to logistics can mean the difference between a successful or unsuccesful demonstration.

Police and other governmental agencies

The police should be notified ahead of time about the time and place of each demonstration. Try to develop a good relationship with the police who will be monitoring the demonstration. In some police departments there is a special division or subdivision called the civil disobedience or community relations department. When the police arrive at a demonstration, explain the purpose of the demonstration to them. Explain that you are there as a nonviolent witness and will do whatever you can to see that the atmosphere remains nonviolent. Offer the police food, tea and coffee. Try to engage in a dialogue with them after the demonstration is over about how they experienced the demonstration and evaluate the demonstration with them if you can.

Campaign coordinator committee

The campaign coordinators facilitate the entire campaign by making sure that various tasks are being worked on, and facilitating communication between the various task groups.

Demonstration Facilitators

This is the team which oversees a particular demonstration by coordinating the efforts of the various task groups, seeing that meetings and activities get started on time, acting as spokespeople to the press and, in some cases, acting as part of the negotiating team. A demonstration is a production and needs "directors" who are looking at the whole thing to make certain everything is going according to plan and to spot problems and correct them quickly. The campaign facilitator should be able to make quick decisions about tactics and strategy, but the entire group usually makes the final decision on any major changes in action. Facilitators are not "leaders," but people who have chosen to give their

time, energy and whatever special talents they may have to work toward nonviolent change. Some of the facilitators should be rotated after each demonstration, but each team should include both experienced and unexperienced people. Three to five facilitators is an adequate number. Some of the facilitators should *not* get arrested.

Relations with other groups

Maintain contacts with other social change groups. Let them know what you are doing, how they might participate in or support your efforts. Grass roots groups should try to make contact with unions, church groups, professional organizations and schools to explain what you are doing and why. Invite members from other organizations to your group meetings, particularly members of grass roots organizations who might have expertise in a particular area such as transportation bans or initiative petitions.

Nonviolent training, campaign training, and stewards for demonstration

Give *nonviolent training* to the demonstrators. Hold sessions before the day of the demonstration and give last minute training or instructions the day of the demonstration. The *stewards* (or "marshals") help the demonstration facilitators maintain a nonviolent and orderly demonstration. Their job is to help facilitate the demonstration. Sometimes they may walk alongside the demonstration. They should be clearly designated, perhaps by a colored cloth on their arms. They should help demonstrators maintain a reasonable formation, and be trained to maintain nonviolence even if the group is attacked or harrassed. A steward should be warm, friendly, humble, but firm and courageous when necessary. The steward should avoid yelling, pushing, demanding, ordering or bossing demonstrators.

Office

The campaign should have an office, an address, phone number and a special number where callers can obtain information about the demonstration. It would be a good idea to have someone at this number during the demonstration to advise "last minute" demonstrators how to procede.

Civil Disobedience

Special concern and problems focus around arrest. Lawyers are needed. An approach should be agreed upon about arrests. The number of people to be arrested should be decided before the demonstration. Usually only a small number of people are arrested so that you do not use up all your energies and finances on one demonstration. Work closely with the legal support team and legal coordinator to understand the judicial system and what to expect after you have been arrested. Be prepared for jail. Try to avoid overidealizing civil disobedience.

Child care

Many people would like to attend demonstrations but cannot because they have young children. Grass roots groups should work with other local and regional groups to assure that day care is provided for parents who wish to demonstrate. Young children should not attend civil disobedience demonstrations because child welfare workers are sometimes at these demonstrations and parents may lose custody of their children following arrest.

SUSTAINING A NONVIOLENT CAMPAIGN (SOME THEORETICAL AND PRACTICAL NOTES)

People's movements require peoples' participation and democratic decision making. Sharing responsibilities helps everyone get involved and gives everyone more experience. Rotating leadership and sharing responsibilities helps maintain group morale and limits the chances of any one person gaining too much power. We have all been raised to compete, but nonviolent demonstrations have shown that we can learn to cooperate. Because a nonviolent demonstration requires a good deal of discipline and adherence to previously established group decisions, certain policies can not be changed once the demonstration has started. In many ways the eyes of the world are on us when we participate in a nonviolent demonstration and we can not afford to "do our own thing" in this setting.

Esprit de corps
A successful campaign requires a high, positive, friendly spirit. The physical and spiritual needs of the participants must be carefully tended. People's needs should come as a high priority at all times. The meeting of physical needs includes food, drink, toilet facilities, warmth, sleep, medical care, etc. Meeting spiritual needs and keeping up morale may include singing, humor, openness, honesty, democratic participation, good communication, informality, etc. A strong structure is needed, but one that allows for and meets these needs--a humanizing structure and one that allows for doing fun, even silly, things.

On-going evaluations
Evaluations should be made after each meeting and each event by the entire group.

Openness
Openness about our ideas and actions is extremely important. It is part of democracy, truthfulness, maintaining good morale, maximizing our power, undercutting the power of agent provocateurs, and helping build the movement. Secrecy on the other hand does just the opposite. It breeds paranoia, suspicion, and rapidly escalates into a time consuming and mutually destructive drama. It causes more secrecy and often alienates potential converts to our cause.

Overcoming friendly and logical opposition
Every successful peoples' action was logically ill-conceived and ill-timed and in the wrong place with the wrong method-- or so people thought at first. We should listen respectfully when people say our campaigns won't work and then go right ahead and prove them wrong.

Nonviolent discipline
Nonviolent discipline must be absolutely maintained. Any acts of violence by participants will undercut our effectiveness. Violence simply does not mix with a nonviolent campaign. Any violence will most likely be photographed for newspapers or shown on television to discredit the entire demonstration. To maintain nonviolence we need to train participants well, have a discipline statement that everyone agrees to, and stewards and facilitators who are experienced in nonviolent action. During tense times we may need to hold hands, sit down, sing, form affinity groups. If major violence is done by the demonstrators we should stop all

Police mass at Brokdorf in Northern Germany to repel anti-nuclear demonstrators.

planned activities and make it clear that this is *not* what we believe in. Work parties should repair damages, pay for repairs and clean up any area that has been trashed. Special acts of contrition may be in order such as a silent march, apologies to anyone who's been injured, or an offer of reparations for damages in the form of work parties. Evaluation and new preparations are needed to make sure the incident is not repeated.

Utilizing existing organizations

Although some established organizations will be unable to join our campaigns when they first begin, they may be helpful by making their facilities or resources available, donating staff, spreading the word, etc. As the campaign develops, they might eventually take on more aspects of the campaign.

Civil disobedience

Civil disobedience is a two-edged tool. It can be a powerful force or it can be harmful to a campaign. Its advantages are that arrests, if done for the right reasons, can be inspirational. Sometimes arrests are needed to maintain our civil liberties and our right to act. On the other hand, civil disobedience can hurt a campaign if done for the wrong reasons and with the wrong attitude. There is nothing *intrinsically* good about being arrested. Civil disobedience is an act of bravery only when it is done in the right way and for the right reasons.

Time perspective

Don't expect miracles overnight. One problem we encounter as Americans is the desire to achieve things overnight. If we are to dedicate ourselves to nonviolent struggle for social change we should realize that just one campaign may take years. Many of Gandhi's campaigns lasted years.

Ending a campaign

A related question is when to end a campaign. Don't keep a campaign going without reason, i.e. when the momentum has been lost. Direct action campaigns should last two to five months, and then you can regroup, consider your actions and evaluate your progress. A movement is really a series of campaigns which continue until the objective (voting rights, the end of fighting in Vietnam, an end to nuclear power) is reached. In every campaign people will get "burned out" and need time to regain physical and emotional strength. Rest periods are necessary if you wish to be efficient participants in nonviolent struggle.

ORGANIZING A PARTICULAR DEMONSTRATION

Before the day of the demonstration

The scenario for the day is decided upon by the entire group with contingency plans.

People volunteer to do all the necessary tasks and begin work on them.

The demonstration facilitators evaluate progress, see that tasks are being completed, and consult with the legal coordinators to see whether the checklists have been completed (see LEGAL COORDINATOR'S CHECKLISTS) and that the legal support committee is aware of all legal strategies the group may use.

Transportation meeting place

If the demonstration is out of town or in a distant neighborhood, pick an early meeting place where people can congregate at a designated time so transportation can be coordinated and provided for those who need it.

Rumors, Rumors, Rumors

An information overload often occurs during demonstrations. Much of it is conflicting and disturbing. Some things you might hear during a demonstration are: "The demonstration's been called off." "A bomb threat has been made." "They've given in." It is important that we be prepared for such rumors and not act on any dramatic information until it has been investigated. In many cases opponents of a demonstration will call in false information to the press or to grass roots organizations and we must be careful not to be misled. We should also avoid starting or perpetuating rumors.

Notify police about time and place of demonstration ahead of time.

Pre-demonstration assembly and meeting

Every demonstration should assemble some distance (1 to 10 blocks) from the demonstration site. Then march to the site. The bigger the confrontation and the more people you have, the further away the congregation point should be. The assembly site should be a building with comfort facilities. Many churches offer their facilities to demonstrators and provide various types of support for a nonviolent demonstration. Placing the assembly point some distance from the demonstration site accomplishes three things: It enables demonstrators to collect their energies, make last minute changes in plans, and see that small children are taken care of. It provides a setting for the march to the demonstration, which builds drama as the demonstrators approach the site. It enables demonstrators to make contact with one another before the demonstration. *A pre-demonstration meeting* should be held at the assembly site for all demonstrators. Last minute training and instructions should be given at this time, plans and contingencies should be reviewed, discipline commitments should be reviewed and any questions regarding discipline answered, solidarity statements should be made, a general statement about the demonstration given.

March

March to the place of the confrontation or demonstration. Take your time, don't panic, sing.

Violence

In case of violence or pending violence, sit down, hold hands, don't run or panic, sing, follow stewards' and facilitators' instructions, maintain a disciplined formation, and utilize the exercises you practiced before the demonstration.

Preventing police from stopping a demonstration

Had demonstrators obeyed police orders, most demonstrations would have been blocked. Police are often the recipients of false information from government agencies, business organizations or other opponents of a demonstration. Often the police are frightened and overreact to the demonstrators. They will also use various approaches, ranging from extremely friendly to hostile, to convince you that what you are doing is illegal. Before the demonstration your group should examine the various laws you may be violating so that you know exactly which laws you are *intentionally* breaking. This will help you avoid confusion and enable you to respond to a police officer's attempt to provoke you into quitting the demonstration.

Do not block public throughways (unless deliberately doing a civil disobedience blockade.)
When picketing be sure to leave room for pedestrians on the sidewalk.

Communications
Use short information meetings *during the demonstration* to keep participants informed about progress and to examine strategy to determine its effectiveness. Runners can visit the different areas of the site with information about the overall picture and news from the various affinity groups in different parts of the site. Facilitators should roam between areas, but at least one facilitator should be permanently at each area.

Ending the demonstration
Before the demonstration begins, your group should agree on a time and place to meet when it is over. Much confusion can result when the demonstration ends and people can not find rides, children, etc. This confusion can be avoided by agreeing on a plan of action some time before the demonstration.

Evaluation
The evaluation could be written on flip cards in three columns: positives, negatives and how to do things better the next time.

A CHECKLIST FOR DEMONSTRATORS

A mass demonstration against nuclear power requires organization, discipline, and a knowledge of the possible consequences of one's act by each participant. The planning for a mass occupation begins months in advance so there will be a minimum of confusion when the time arrives to commit civil disobedience, and later when demonstrators are being arrested, booked, paying or waiving bail and going to jail. One of the most important facilitators in any such action is the legal support coordinator who serves as a liaison between demonstrators and the support committee of volunteer lawyers and paraprofessionals. In the weeks preceding the demonstration or occupation, the coordinator (or coordinators if the group is large) will distribute the following checklists to each potential demonstrator. This information is vital to the coordinator and the support team as an indicator of how well prepared each participant and affinity group* is for the demonstration. The information will also be used by the coordinator to keep track of who is in jail and where, and to help those arrested go through legal processing with a minimum of confusion and, hopefully, little of the despair or depression that can accompany this procedure. The coordinator will make sure that the legal support team has a copy of the checklists and will keep a copy of each list.

The checklists were designed by the Clamshell Legal Support Committee which has done considerable research into the legal aspects of civil disobedience.

SAMPLE LEGAL COORDINATOR'S CHECKLIST

You should compile immediately a thorough list of the following information from each potential occupier in your group:

> name of occupier
> home address
> home phone
> work place
> work phone
> age
> sex
> nature and place of employment
> name of occupier's support person (i.e. person with bail and responsibility for occupier's welfare)
> time occupier needs to bail out by
> list of occupier's special needs
> prior arrests or convictions
> name of attorney to contact (if any)
> occupier's plans re: hunger strike and/or non-cooperation

A copy of the above information should be available to your group's office and you should keep one copy in your possession.

*An affinity group is a small number of people who have developed a mutual trust and who offer one another support during a civil disobedience demonstration. Forming affinity groups is one way of limiting the chances of disruption by provocateurs, keeping the demonstration orderly, and facilitating the work of the legal coordinator.

The following checklist should be of some help to you in performing your legal support function.

☐ I have compiled the data needed on all potential occupiers in my group.

☐ This data has been provided to the (your local group's address or the address of a regional office) and I have kept a copy.

☐ I have contacted the clerk of the appropriate court to determine how bail must be paid. It must be paid as follows:

☐ Affinity group has discussed what legal research, etc., it wants from Legal Support. It wants the following information:
By what date:

☐ Affinity group has discussed the possibility of arranging for an attorney to be available to work with the affinity group re: this action.

☐ If it is considered desirable by affinity group, I have established informal contact with clerk of the court, chief of police, chief of security, etc.

☐ Affinity group has discussed the following:

> ☐ Bail solidarity
> ☐ Individual non-cooperation
> ☐ Group non-cooperation/hunger strike etc.
> ☐ Bail reduction hearing and its objective.

CHECKLIST FOR CIVIL DISOBEDIENCE PARTICIPANTS

PREPARED BY THE CLAMSHELL ALLIANCE LEGAL SUPPORT COMMITTEE

PERSONAL LIFE AND POSSESSIONS

☐ All personal possessions have been given to my support person in a labeled bag.

☐ I understand that I may not be able to get my medicine while in jail.

☐ I have called anyone who needs to know my whereabouts for the next few days after arrest or longer.

☐ I understand that there is no guarantee that I will be allowed a phone call soon after being arrested.

☐ I understand that carrying anything that may be considered a weapon could subject me to further charges.

BAIL

☐ I know what bail is usually set for this offense.

☐ I understand that bail may be set at a different amount for me and that it might not be reduced after a hearing.

☐ I know what form bail money may take in the county I may be arrested in.

☐ I know what days and times of day bailing out is possible in the county I may be arrested in.

☐ I understand that all actions and words of police, jailers, and judges may have some effect on my case and I will take note of what occurs.

ARRAIGNMENT

☐ I know what a typical arraignment is like in _____ County.

☐ I have thought about what I will say at arraignment and how I will say it.

☐ I understand the meaning of pleas such as guilty, not guilty, no contest (nolo contendere), and a creative plea (i.e. I plea for my children)

☐ I know the possible arraignment schedule for _____ County.

☐ I know what personal recognizance is.

☐ I know that conditions can be put on my release under personal recognizance.

☐ I understand that there is the possibility that I will be thrown out of the courtroom or jail if I refuse personal recognizance.

☐ I have discussed jail/bail solidarity with my affinity group and other affinity groups.

☐ I understand the difficulties in maintaining solidarity in jail.

POST ARRAIGNMENT PRE TRIAL IN JAIL

☐ I understand a trial may be set for me at a date several weeks or months after the arrest.

☐ I know what a bail reduction hearing is.

☐ I have seen examples of bail reduction motions.

☐ I know how to file a bail reduction motion.

☐ I understand that I may be separated from my affinity group and placed in a private cell or with members of the general prison population after arraignment.

☐ I know the location of jails I may be taken to if I do not bail out before arraignment or trial.

LEGAL SUPPORT

☐ I understand that no member of the legal support team has committed himself/herself to represent me in any capacity.

☐ I understand that there may be no member of the legal support team at the waive action, jail or arraignment.

MISCELLANEOUS

☐ I know the possible penalties for the acts I may take and am prepared to fully accept the consequences.

☐ I understand the penalties I may incur if I violate my personal recognizance or bail agreement.

☐ I know what pro se (self-representation) involves.

☐ If I intend to go pro se I have attended a pro se workshop and/or know what I am going to do.

☐ I understand what contempt of court is and its penalties.

☐ I know whether I am eligible for court appointed counsel.

☐ I understand that if I receive court appointed counsel I may have to reimburse the court at a later date (depends on state).

☐ I understand the arrest and booking procedures of _____ County.

☐ I know the history of occupations and legal struggles around the _____ nuke.

☐ I have had non-violence training and know the _____ (alliance or group) guidelines.

☐ (if a juvenile)--I know that I may be treated differently than adults.

☐ I have spoken with someone who has committed C.D. in _____ County.

☐ I understand that men and women will be separated in jail.

☐ I have read an occupiers handbook.

☐ I understand that I may be separated from my affinity group or companions at any time.

☐ I will wear clothing in anticipation of a cold jail cell.

INDIVIDUAL GROUP MEMBER'S SUPPORT INFORMATION

We have selected a legal coordinator (someone who will not be arrested).

She/he is _____

Her/his phone number before occupation

_____days

_____nights

Phone number during occupation

_____days

_____nights

Phone number after occupation

_____days

_____nights

Legal coordinator's mailing address:

I have provided legal coordinator with the following information:

My name, age, sex, mailing address, phone, special needs, person to be contacted in case of emergency and any and all other logical pieces of information that a legal coordinator should have.

CIVIL DISOBEDIENCE AND LEGAL STRATEGY

By Scott Kennedy

Scott Kennedy is on the staff of the Resource Center for Nonviolence in Santa Cruz and is also active in People for a Nuclear Free Future, a member group of the Abalone Alliance. In this article, taken from the June, 1979 issue of WIN magazine, Mr. Kennedy examines some of the difficulties demonstrators face when being processed through the courts, and argues that if we are arrested and brought to trial we should demonstrate the courage of our convictions by avoiding "redundant courtroom strategies" and, when necessary, spending time in jail.

The *China Syndrome* and Three Mile Island accident, as media and public events, fell on ground made fertile by several years of hard work and imaginative actions by the anti-nuclear movement. Both events gave an unexpected boost to a rising swell of public opinion against nuclear power and weapons.

The anti-nuclear movement has definitely entered a new phase. All aspects of our work must undergo renewed evaluation as we plot our future course of action. There is a wide spectrum of activity endeavoring to end nuclear power and weapons. The distinctive contribution of the grassroots anti-nuclear network is the commitment to nonviolent direct action, including civil disobedience. It is our greatest contribution to the larger movement of which we are a part. Assuming that civil disobedience continues to play a significant role in this movement, it is important to look at our legal strategy.

We are often not as thoughtful in preparing our legal strategy as in our preparation for direct action. This is partly because legal terrain is unfamiliar turf. In California, the great distances which people arrested at Diablo Canyon must travel for court dates in San Luis Obispo discourages informed and active participation. Complicated and prolonged legal processes tax the scant resources of the movement: our finances, volunteer energies, public credibility, media attention, citizen initiative and involvement. Despite *our* wishes and *their* best intentions, reliance on lawyers casts people's fate into the hands of a few experts in a process scarcely understood--a dynamic at odds with the expressed objectives and style of a nonviolent movement.

The legal fate of those arrested at Diablo Canyon in 1977 is still undecided nearly two years later. Most occupiers do not know what, if any, their sentences will be. In the aftermath of Diablo '77, those going through the courts felt abandoned by others in the Alliance at critical junctures. There is a great reluctance to join in further direct actions.

It is not civil disobedience but the legal process about which people have reservations. Understandably, people are concerned about added legal jeopardy for repeat offenders. Others have found their experience of legal affairs characterized by confusion, powerlessness, isolation, non-communication and frustration.

In 1977, legal maneuvers often aimed at securing acquittals or dismissal of charges despite the merit or relevance of the arguments employed to achieve these ends. An early defense even tried to prove PG&E did not hold legal title to the land on which people trespassed! Some defendants contested the fact of being arrested, claimed they were not given adequate warning before arrest, or offered other defenses quite remarkable for people who willingly undertook to break the law.

The few who pled guilty or no contest came under strong peer pressure to join a unified legal defense with professional counsel, technical defenses, etc.

In the wake of Diablo 1978 and the arrest of nearly 500, legal strategies of the Abalone have again outdistanced most people's interest, time, energy and financial resources. Dissatisfaction with the legal strategy is widespread. The tenfold increase in arrests multiplies the questions relating to legal strategy.

For one thing, it reveals the inadequacy of the Abalone's training. The generally excellent training for civil disobedience scarcely addressed what would happen after arrest: incarceration, legal defense, pleas, and going to jail. The Direct Action Handbook for Diablo '78 described jail as a "meet great people kind of time," as though it were (to borrow from the lexicon of movement jargon) a "light and lively."

Legal briefings gave many the impression that the Abalone was providing legal defense and guaranteeing minimal legal consequences, the first beyond our means and the second beyond our ability. It gave the impression that peer pressure solidarity alone could determine the circumstance of people's confinement or legal consequences. Too many people thought that because of the large number arrested they would be spending little or no time in jail and given minimal fines.

The acceptance of serious consequences from civil disobedience must somehow be communicated to people. The relative strengths of various approaches to pleas and legal defense should be as rigorously examined and thoughtfully understood as other aspects of our movement (organization, decision-making or direct action).

It boils down fundamentally to one's views of nonviolent direct action and civil disobedience and the nature of the power of these tactics.

The power of nonviolence is not manifested by eloquent, technical, or redundant courtroom strategies. Its strength is based on the preparedness to accept the consequences of our action, including suffering, rather than inflicting it on others. The sincerity of our opposition to nuclear power and weapons is evident and encourages others to escalate their own activity for the same cause. This power of civil disobedience is not served when charges are dismissed or acquittal won on grounds unrelated to the cause for the civil disobedience. Strength is evident in our readiness to act in spite of intimidation by the state, not by winning according to *its* rules. The true test is to persuade enough people of the justice of our cause to radically alter the political and social climate for all.

With civil disobedience, our best propaganda is our free activity which accepts the possibility of "suffering" --arrest, detainment, jail or worse.

Direct action despite legal threats and repressive measures makes evident to the public and to our adversaries the sincerity of our convictions, the urgency and seriousness of the issues, the growth and strength of our movement, and the necessity for public discussion, debate and decision.

Large numbers of people being arrested and going to jail seems to have a dramatic impact on the public. Acquittal or dismissal of charges on points tangential to the issue itself do not clarify what is at stake or, in the long run, further the anti-nuclear movement. It does not prepare people to engage in militant nonviolent action which involves accepting personal jeopardy in collective pursuit of our goals.

Courtroom trials are often defended as a way to speak to people and raise issues. At the same time, energy and funds committed to ongoing public outreach, education and direct action would reach far more than the 12 persons on a jury and whatever coverage the media deems fit to provide.

This is especially true of a series of trials, which the media will likely grow tired of and treat cynically if at all. A judge can (and usually will) arbitrarily rule out of order and inadmissible any reference to the issues which prompted civil disobedience: defenses of necessity, international law, competing harms, etc. One can scarcely imagine a forum where we can exercise less initiative and control of events than in the adversary setting of the courtroom. As Peter Klotz-Chamberlin wrote from San Luis Obispo County Jail, during a two-month sentence for Diablo '78, "Courts lock us up just like the jails do--only in a more insidious way." More insidious because we don't realize it is happening.

Some strategies seek to manipulate the legal system, placing great stress on "solidarity" and strength in numbers--unanimous action, at the expense of individuals acting in concert with their best sense. When the movement uses large numbers as a way of exercising power in the courts, we succumb to the very methods we seek to change.

Historically, nonviolent action has viewed its power as emanating from the truth of the cause, the clarity of its action, and not from the treatment of individuals by the courts. Having chosen civil disobedience as part of a larger strategy, dignity and courage are communicated by the acceptance of legal consequences. This in turn exerts a tremendous moral force on the state, clarifying issues and eroding public support for those using the law to prop up nuclear technology.

Guidelines for Action

The same criteria by which we judge our direct action, organizing structures and decision-making should be applied to our legal strategies. Respect for individual choice, attention to means as well as ends, self-governance and participatory forms should be highly valued. Any legal maneuver not clearly related to stated objectives and operating principles should be avoided.

In many ways, the ideal scenario after arrest has all persons in jail as soon after the action as possible. This creates a greater likelihood of maintaining the cohesion of the group, sustaining energies, and reinforcing commitment to group struggle. Continuity between the action and jail would maintain momentum for the action instead of dissipating it. This also places the state in the dilemma of having to justify their choice to arrest and incarcerate protestors. It focuses public attention on the action itself, rather than on peripheral issues. After jail, people can begin planning the next phase of their work. The work does not need to stop while in jail, as those given lengthy sentences for Diablo '78 discovered during their time of incarceration.

Based on recent experience with resistence to Diablo Canyon, the most clear and powerful action may well be pleading "no contest" to charges, requesting sentencing and imposition of sentence as soon as possible. In fact, this year's participants in a blockade of Diablo are being encouraged to plead "no contest" and be sentenced immediately. Declarations about motivation and the larger political context for direct action can be made at sentencing.

No action should be changed or any strategy undertaken out of fear of legal jeopardy. When the first people sentenced to 15 days, a fine and two year probation for Diablo '78 refused the terms of probation, they were given six months in jail and a $500 fine, the maximum for a misdemeanor. As Sam Tyson, who served 62 days in jail for Diablo '78 wrote, "This set the limits within which things could happen, a stark reality for many. The idea that there won't be casualties in nonviolent direct action is naive in the extreme."

Some other guidelines can be suggested:

(a) Encourage non-payment of fines. Despite hundreds of "mandatory" fines for Diablo, totalling nearly a quarter of a million dollars, the state has been unable or unwilling to collect very much of it.

(b) If people wish to "put nuclear power (or weapons) on trial," or to test violations of international law, a single such trial can be undertaken and done well. The Abalone negotiated for a representative trial of 20 defendants after Diablo '78. This allowed for consolidation of efforts, pooling attorneys, and minimizing expenses while maximizing public awareness of the trial. Even so, it took several months of people's time and energies, and cost thousands of dollars. And it is questionable whether there was enough support outside of San Luis Obispo County to assure sufficient public interest and media attention to justify the trial.

If defenses of necessity or defense based on international law, or evidence related to the criminality of nuclear weapons and power are disallowed by the court, the trial strategy can be abandoned and pleas of no contest entered.

(c) Those committed to nonviolent action leading to trial can extend militant tactics into the courtroom. Why become passive and accept the limits of behavior acceptable to the state just because you are in court? When anti-nuclear jurors are dismissed, others could walk to take their places in the jury box, insisting that defendants have a right to a trial by their peers.

(d) Some defendants may choose simply to remain silent in court before their accusers.

(e) Those who want trials can defend themselves. They can stipulate to the circumstances of their arrest to avoid a useless focus on technicalities which avoids the real reason for the action. Defending oneself is an experiment in self-governance and can be an empowering experience, as contrasted with most people's experiences of being defended by lawyers. (I am of the opinion that lawyers committed to the anti-nuclear movement would probably have a far greater impact by being arrested with us than by volunteering their services as defense counsel.)

(f) In order to keep things in perspective, a ratio of expenditures can be agreed to, setting aside a few dollars to implement the next action for every dollar spent on legal strategies for the former action. This keeps the focus on the future and ongoing work.

(Obviously these suggestions apply to situations in which we are largely responsible for generating arrests. Most civil disobedience is meticulous in its planning and clear in its willingness to risk arrest. If the state initiates legal action, however, as in the Chicago and Harrisburg conspiracy indictments, or charges people with crimes that were not committed, then the movement should feel free to muster its resources in the most powerful fashion to defend against such an attack.)

Our objectives and commitments to a nuclear free future are best communicated, not by proclamations and declarations, but by our action. Devotion to the nonviolent struggle against nuclear weapons and power is our greatest resource. We must draw on it with great care. Once civil disobedience has been agreed to as an appropriate response, a thoughtful legal strategy must be developed. Complicated and protracted legal entanglements serve to draw out, exhaust and distract people in the movement and to confuse and confound the public. Trials spread out over months or even years reduce solidarity among defendants, lose the immediacy in the public's mind linking legal consequences to the action itself, and increase the likelihood of people doing jail time in isolation.

Nonviolent action is the linchpin of all our other important efforts at public education and outreach, coalition building, creating alternatives, and developing a broadbased and vital movement against nuclear power and weapons. As the movement gains strength, the weight of the courts may be increasingly felt in a wave of reaction and repression. We should evaluate our approaches and understand those which strengthen us for such eventualities. Our forerunners in nonviolent action have shown that civil disobedience is most powerful when its practitioners voluntarily accept legal and other consequences without defensiveness, without attempts to mitigate or avoid them, and without fear.

DEFENDING ONESELF WITHOUT A LAWYER

For participants in civil disobedience who wish to defend themselves (pro se defense) the Anti-Nuclear Legal Project in Boston has written a legal guide. Copies of the handbook can be acquired by writing: Anti-Nuclear Legal Project, c/o Massachusetts Lawyers Guild—Room 1011, 120 Boylston Street, Boston, Ma. 02116. The cost is $2.50 each.

CONSIDERATION OF SELF-REPRESENTATION

The Constitution gives you the right to represent yourself. The right is founded on the understanding that someone else may not say quite what you want said in your behalf, or may not say it in the way you want it said. You therefore cannot be forced to let someone speak for you in so important a matter as a criminal trial.

Trials resulting from civil disobedience at nuclear power plant sites seem particularly suited to unearthing the reasons behind and the possibilities for self-representation. What need is there for a lawyer to explain to a jury an occupier's motivation for blocking entrance to a site? Can't each person explain more compellingly in her/his words? Why water down a political act of civil disobedience with a lot of legalistic mumbo-jumbo? Why let the application of the energizing ideas contained in the philosophy of non-violent civil disobedience stop with the arrests? Why not continue to apply these ideas all the way through the trial?

While we encourage all occupiers to seriously consider the option of self-representation, we recognize that for some it may not be suitable. You should speak to people who have represented themselves and watch a trial or two before you decide.

DECIDING WHETHER TO HAVE A LAWYER

This handbook is based on the assumption that you intend to represent yourself. It is, of course, possible for you to hire a lawyer, find one who will volunteer his or her time, or ask the judge to appoint one for you. You should make sure that any lawyer you involve in your trial understands and shares your political perspective. An unsympathetic lawyer can be worse than no lawyer at all.

There are many personal and political reasons why one would choose to conduct a pro se defense. The following is by Court Dorsey, a member of the Clamshell Alliance, who was charged with trespassing at the Seabrook, New Hampshire nuke site and went all the way to the New Hampshire

Supreme Court pro se, with some help from lawyers and law students. He was so outstanding in his oral arguments that when he finished the whole courtroom (except for the judges) burst into applause:

"Many Clams choose to defend themselves at trial. Faced with what they see as an inherently violent technology, they have chosen direct action as a necessary means of opposing that violence which enjoys the sanction of the prevailing legislative and judicial powers. They feel that a direct personal appeal to the beings of conscience within the jury members, the judge, the members of the press, and all those present in the Courtroom is the truest extension of that direct action, and the means which best preserves the integrity of their moral convictions. They have sought to indict the crime of nuclear power through direct action and wish to address THAT indictment in the Courtroom, rather than be diverted into legal or technical matters of the lettered law represented by the charges brought against THEM.

"The pro se method of Courtroom defense has several advantages in its ability to enhance the political defendant's true concerns. In their ignorance of proper and conventional Courtroom processes, they enjoy a greater latitude in addressing what, for them, are real issues. For example, they may ask prosecution witnesses if they have children and are concerned with the health of children, cutting directly to the heart of the matter. An attorney is far less likely to enjoy such latitude: being familar with ordinary Courtroom practices, the judge and the prosecution can bridle him or her effectively into compliance with the narrow concerns of the law. But the jury member, like the pro se defendant, is lost in the world of legal technicality. If the judge or prosecutor is harsh with the defendant, jury sympathy may be evoked for the well-meaning defendant who is as lost as they. Consequently, judge and prosecutor are likely to curb the energetic enforcement of their legal powers, giving the defendant more room to focus on the real issues of concern. And should the officers of the Court have a vested interest in maintaining an ordinary process of "justice" which they have mastered, their accomplished manipulation of that process is upset by the categorically irregular behavior of the pro se defendant which is allowed in view of the defendant's "ignorance," leaving the pro se person more direct access to the beings of conscience whom she/he seeks to arouse and address. Also, you can get away with a lot more than a lawyer in cross-examination—and this means getting to say things that a jury might not ordinarily hear—you, of course, being well-meaning and naive about courtroom procedures.

"Sympathy is also aroused by the drama of the lone defendant in the face of the overpowering intimidation of the State and Court. One speaks one's own case, with one's own emotion, simply, without the fog of technical distractions, revealing a human concern. In conventional trials, the defendant plays a helpless, passive role, with little opportunity to reveal that one is a human being and not a "villain." In pro se defense, the defendant stands forth in the hush that comes when the one tried speaks for him or herself, and in the hush the voice of conscience can be heard. This sense of drama also appeals to and intrigues the media, and the defendant's statement of

nuclear violence continues to be carried beyond the courtroom, to the public forum where it rightly belongs. So even if the defendant can't overcome the technical instruction of the Court to win acquittal, s/he may "win" the case in those who have been moved.

"It is in the technical aspects of their defenses that the pro se defendants find the greatest disadvantages, and it is here that brothers and sisters with legal expertise can be of most assistance. One of the only technical advantages of the pro se defense is the ability of the defendants to address the jury in opening and closing statements without having to subject themselves to cross-examination. Legal counsel can be very helpful to pro se defendants, when such help is desired. Advice on the filing of motions before and during the trial, advice on jury selection, helping the defendant to grasp the overall "game plan" of the trial and some of the rules and important concepts, from the importance of noting exceptions (*very important* if you want to appeal later, in some states a prerequisite to appeal) to jury nullification—all this can be extremely helpful. The presence of friendly legal counsel nearby, if not at the counsel table itself, is an encouragement and a security to a stranger in a strange land."

AN INTERVIEW WITH TWO ACTIVISTS

By Fred Wilcox

David Greenwald, Wendy Foreman and their son, Abe, live in Bucks County, Pennsylvania. David and Wendy are practising psychotherapists. Because they live only two hours from Harrisburg, Pennsylvania, they felt extremely vulnerable during the near meltdown in March, 1979. Following the accident at Three Mile Island, they joined the anti-nuclear movement by working to oppose construction of a nuclear plant at Limerick, Pennsylvania— a short distance from their home.

On June 4, 1979, Wendy and David, along with members of their grass roots group, the Clean Energy Collective, climbed the fence that surrounds the Limerick plant. As thousands of Americans have done over the past five years, they had decided to express their opposition to nuclear power through an act of civil disobedience. For climbing the fence at Limerick, they were arrested, tried and fined. Some members of their group have been sentenced to jail for refusing to pay their fine. When I interviewed David and Wendy, David was waiting to hear if he would be sent to jail. Nearly four months pregnant with their second child, Wendy had decided to pay the fine.

Wendy and David's experience is fairly typical of people who commit civil disobedience for the first time. During the demonstration, they feel happy. Their sense of purpose gives them the strength to face police lines and imminent arrest. But during the arraignment and incarceration which follow arrest, it is easy to become discouraged. For those facing the judicial process for the first time, the delays, little humiliations, denials of motions and insensitive judges are a bureaucratic anti-climax to the exultation of occupying a plant site.

INTERVIEW WITH DAVID GREENWALD AND WENDY FOREMAN

Q: Why did you choose civil disobedience as a tactic for opposing nuclear power instead of working within the system?

Wendy: We do work within the system. That's what we do every week with our Clean Energy Collective. It's not a question of one or the other for us.

David: I committed civil disobedience at Limerick because of the weekend of terror that Three Mile Island subjected all of us to. I wanted to do something dramatic, something to purge the feeling of helplessness I experienced during the accident at TMI. Climbing the fence at Limerick and getting arrested was just my way of letting those people know that I was ready to do anything to keep that from happening again.

Q: Did you receive any training before the action?

Wendy: Yes, we were trained by a man who had been trained by Keystone Organizers. The organizers of the Limerick action had a very definite rule: absolutely no one would be allowed to commit C.D. on that day who had not been trained. Some people showed up at the last moment and said, 'Well, I've read Gandhi, or I've done this or that. . .' But the organizers had to protect the action against provocateurs or people who might do something stupid. I thought the organizers' refusal was legitimate because if people were committed enough they would spend a few hours in training.

Q: Was your son with you when you climbed the fence at Limerick?

Wendy: Abe could not be with us because he would have gotten into the hands of the child welfare bureau and been taken from us for an unknown length of time. A woman was there from the welfare department just waiting for the opportunity to accuse someone of child neglect. We did take Abe to the demonstration in Washington last April, but Limerick was a disciplined action which required the participants to be fully cognizant of their individual acts of civil disobedience.

Q: Now that some time has passed since you committed civil disobedience, how do you feel about your action? Would you do it again?

Wendy: It was a little disappointing but I would do it again if I were sure it would have a little more impact. If it were planned with a much larger group of people, or if we could be certain we would receive a jury trial. By not giving us a jury trial, trying us separately and reducing the charge, the judicial system effectively diluted the action.

Q: Were you prepared for what happened in court?

Wendy: We were totally unprepared. The subject never even came up during the training sessions.

David: Watching the judge go to lunch with the head of Limerick's security and the prosecuting attorney certainly wasn't my idea of Perry Mason.

Wendy: The township building where we were tried was built in part by the utility that owns the Limerick Plant, and behind the judge was a large logo with a picture of the nuke unit. We all had a pretty good laugh over that.

Q: What was the judge's attitude toward you?

David: He refused to take us seriously. On the day I read my statement he kept rubbing his eyes, yawning, trying not to hear.

Wendy: Our trial was unique in several ways. First because we were tried by district justices, who are justices of the peace, elected officials without any formal legal training. Second, we were tried on different days and each day there was a different justice and a different prosecutor, each more obnoxious than the last. When I gave my statement I thanked the judge for the opportunity to address the court, and he said, 'Well, you're quite welcome Madame, and for the opportunity you will pay fifty dollars in fines and twenty-six dollars in court costs.' That's the way it went. But it was inspiring to us to hear some of the Quakers speak and to be together to express our feelings.

Q: Did you have a group defense?

Wendy: We had one plan based on the competing harms approach, but it was denied.

Q: So when you confronted the legal system, you felt frustrated?

Wendy: We were also frustrated during the demonstration. It was raining when we arrived, and the utility had erected this little snow fence just outside the property. We had planned on getting arrested right away, but the snow fence presented a tactical problem. We started to caucus and this went on for hours, until most of the press and support people had gone home. We just couldn't reach a consensus and, even though it was democracy in action, we lost a lot of potential coverage. The PECO security guards were laugh-ing almost hysterically at us running back and forth, huddling in the rain.

David: The real disagreement was between people who wanted to stay there all night to show our resolve and others who thought it was more important to take a more dramatic and symbolic action.

Q: How was the press coverage?

Wendy: It was fairly positive. There were front page headlines and a photograph of people going over the fence. But when it came to trial, the press seemed to lose interest.

Q: If you should be required to spend time in jail, how will you explain this to your son, Abe?

David: I would say that because I went to a demonstration at Limerick and protested against nuclear power I was arrested. I believe I did the right thing, but since the judge didn't agree, I may have to go to jail for awhile.

Q: As family therapists, what advice would you give people who would like to get involved with the anti-nuclear movement, but say they cannot because of their children?

Wendy: It's because of our children that we *must* get involved.

Q: Some psychologists have referred to civil disobedience as a kind of acting out of an adolescent script. Can you comment on this?

David: That doesn't surprise me at all. Psychologists can call anybody any name they choose and it sounds o.k. But to me if you are legitimately afraid of being annihilated and you choose not to do anything about it, that seems more a hangup than trying to deal with the threat. Then there is the old saw that protest is acting out against daddy. This is an enormous oversimplification of Freudian thought. Actually, it can just as easily be turned around to mean that someone who *doesn't* protest still sees himself or herself as a child and is afraid of losing parental approval. In all the respectable psychological studies of sixties activists, the activists were found to be better adjusted. They had better personal relationships, higher grades, and a higher degree of independence, and fewer authoritarian tendencies.

Q: Do you feel that it's necessary to spend time in jail in order to demonstrate your opposition to nuclear power?

David: I hope not. I feel that people should get involved on any level they can handle. I wouldn't say that everybody should be willing to go to jail. There has to be room for everyone to make a contribution, whether it's writing letters, demonstrating, voting for candidates who are against nuclear power, lobbying. . .there's an entire range of activities, all of which are worthwhile in a mass movement like this.

Civil Disobedience 51

AT THE TRIALS OF THE FENCE JUMPERS

By John Gofman

The following article is from "IRREVY" An IRREVERENT, ILLUSTRATED VIEW OF NUCLEAR POWER by Dr. John Gofman, Professor Emeritus of Medical Physics at the University of California, Berkeley. Dr. Gofman is also a former Associate Director of the Lawrence Livermore (radiation) Laboratories where he conducted research on cancer and chromosomes (1965-72) until the Atomic Energy Commission stopped his work. Today Dr. Gofman is chairman of the Committee for Nuclear Responsibility and one of the nation's most outspoken critics of nuclear power. In this article, he challenges the utilities' right to "commit legalized murder" and calls nonviolent demonstrators the "Paul Reveres of today."

It is my purpose here to examine what I consider some of the most profound threats to freedom and justice which our society has yet witnessed. In order to clarify how such a serious claim comes to be discussed in relation to ways of opposing nuclear power, I must go over a bit of recent history.

THE TOWER-TOPPLING TRIAL

In 1974, a young Amherst graduate and communal farmer, Sam Lovejoy, chose to express his opposition to nuclear power as an abominable threat to humanity by toppling a meteorological tower in Massachusetts, said tower having been constructed to obtain data needed for later construction of a nuclear power plant. He chose to commit his act of civil disobedience on Washington's Birthday. Having toppled the tower, he promptly turned himself into the authorities, finally convinced them that he had indeed toppled the tower and managed to get the authorities to indict him for this apparently dastardly act.

In due course, he came to trial where he served as his own attorney, with the obvious purpose of putting nuclear power on trial. In order to carry through his purpose, he needed expert testimony concerning the hazard of nuclear power to health. He asked me to testify on his behalf, and I agreed that I would.

When it came time for Sam to call me to the witness stand, he stated that he wished to have the hazards of nuclear power explained, and promptly the prosecuting attorney objected that the hazards of nuclear power were irrelevant, since the issue was whether or not he had maliciously destroyed property, not whether nuclear power was dangerous.

The Judge seemed a bit perplexed about ruling on the prosecutor's objection. The course he took was to let Sam conduct a direct examination of me for about 1½ hours, *but without the jury being present.* Then he adjourned the Court for a period while he considered whether the jury should or should not have the opportunity to hear my tes-

timony concerning the health hazards of nuclear power. Finally he returned to the courtroom and announced that he would *not* let the jury hear my testimony; in other words, he sustained the prosecutor's objection.

But the Judge did something extraordinary in addition to denying the jury the opportunity to hear about the hazards of nuclear power. He stated to the Court that in the event that Sam Lovejoy was found guilty, he would not pass sentence but would instead refer the case to the Massachusetts Supreme Court to decide whether he had erred in denying the jury the opportunity to hear and consider the evidence concerning nuclear power in reaching its decision on Sam Lovejoy's guilt or innocence. As it turned out, that issue never came up because the entire charge against Sam was thrown out on a faulty indictment basis.*

There was something about that judge which I thought then was profound. In the four years which have passed I have given the matter a great deal of thought, and I am now certain that my initial impression that this Judge was a serious thinker was indeed correct. Clearly something was troubling the Judge profoundly in reaching his decision to withhold the evidence concerning nuclear power from the jury. A lesser jurist could simply have sustained the prosecutor's objection. A lesser jurist could simply never have made that statement concerning sentencing and the possibility that he had erred. Surely judges are not popular with the Establishment if they do *not* hold property rights as paramount and far more important than human life or health.

I'm glad that our society considers private property rights to be sacred. But also I hold that the right of private property (like the right of free speech) is *forfeited* by people who use it to deny another sacred right, namely someone's inalienable right to *life.* It is ludicrous to suppose that an electric utility which is using (or proposes to use) its property to commit premeditated random *murder* retains all the sacred rights of private property! Likewise, it is ludi-

---

* "Lovejoy's Nuclear War," a one-hour documentary film.

crous to suppose that the Nazi Party, which has not only advocated but also committed genocide, retains the right of free speech. It would make a mockery out of the right to life, to allow either the right of free speech or the right of private property to be used *to deny the right of life* to anyone.

I believe it is interesting and important that the Lovejoy trial was held in New England. I say this because I rather suspect that New England is one of those rare places in the United States where people still have a recollection that there ever existed something called the Declaration of Independence.

Something was indeed bothering that Judge, something which prevented him from a *summary* dismissal of Sam's right to have nuclear power discussed before the jury in its evaluation of his behavior in toppling the tower. The more that I have thought about this episode in the past four years, the more convinced I have become that this Judge recognized that *justice and the law are not synonymous,* and that justice was really of importance. Not many people ever think about this distinction, or even admit that justice and the law are two separate entities.

THE ISSUE: HUMAN RIGHTS

I do not now hold, nor have I held in the past, any opinion one way or the other concerning whether civil disobedience is a good or bad approach to fighting nuclear power. I simply do not know. There is no doubt in my mind that the promotion of nuclear power is itself a criminal act of the worst sort, given the Constitution under which we live. No doubt at all.

The Constitution of the United States does not permit the taking of life without due process of law. Nuclear power, which begins its random murder of citizens of the United States even before the nuclear plant goes into operation, is obviously an infringement of Constitutional rights. Even if there were no Constitutional violation, nuclear power is a violation of natural rights and justice.

There are those in our society who don't realize that justice is a concept which exists quite independently of any laws made by legislatures—even though that independence is commonly acknowledged whenever someone says, "That law is unfair."

Moreover, probably most Americans have forgotten the Ninth Amendment to our Constitution, stating the fact that certain rights inhere naturally to the people whether or not the Consittution spells them out specifically. *The right to life* is certainly one such natural right.

I mentioned above that nuclear power starts to commit murder even before the plant goes into operation. It does so by guaranteeing that people are going to be poisoned for hundreds of thousands of years by radon and its daughter products brought to the surface of the earth in the course of mining the uranium needed to operate the nuclear plants. Had these substances remained in the bowels of the Earth, they would have done no harm.

And the random murder of citizens is further increased by the fact that there are emissions of radioactive substances both in the normal and abnormal operation of the nuclear fuel cycle. Just how large the number of people to be murdered depends upon the outcome of the largest experiment upon humans that evil genius has yet concocted. I say that this experiment is evil in the extreme because it not only kills humans of this generation, but also because it reaches into countless future generations with its lethal effects. That certainly qualifies as the depth of depravity, particularly because future generations may have much better common sense than ours does, and might not elect to be poisoned if they had the opportunity to decide for themselves.

For our purposes here, it is immaterial whether the numbers to be murdered by the operation of a single plant, such as the Palo Verde plant, are to be counted in the tens, the hundreds, or the millions. Murder is murder! We have not yet adopted the position that criminals who kill people have to kill *a certain number* before their action qualifies as a crime.

It may indeed be true that society could reach the decision that it is all right to murder a certain number of people in the course of industrial activities, but that is a profound decision and should be taken only with full cognizance of its true meaning.

LAW VS. JUSTICE

It is said that nuclear power plants can operate *legally* simply because they are licensed to operate by the Nuclear Regulatory Commission. The Nuclear Regulatory Commission is operating *legally* because Congress legislated it into existence to issue such licenses. But what has all this to do with justice and natural rights? Congress has no authority under the Constitution to issue murder licenses. Moreover, Congress could have no such authority, simply because one of the rights protected by the Ninth Amendment is the *natural right to justice and to life.*

That is my opinion, and it would not be altered one whit if there were 100 decisions by the Supreme Court which stated that it is permissible to murder people. There is a higher law.

It amazes me that people don't seem to realize the implications of permitting laws to be passed which violate justice and natural rights. It amazes me since it is so soon after the Nazi Holocaust and the Nuremberg Trials. In Nazi Germany the rulers, as evil a people as one can imagine, wished to carry out a program of genocide. Because of the recognition that people *might* object to such a gross violation of justice and human rights, even the Nazis decided to make the process legal, at least in part, by passing laws which permitted judges to send people to their death with no justification at all other than a Nazi-passed law.

At the Nuremberg Trials, the United States declared that this sham of using "*laws*" to subvert *justice* was a heinous crime, and we meted out severe sentences to judges who had used the Nazi "laws" as a shield for the crimes which they (the judges) committed on the bench.

If the Congress of the United States can permit the Nuclear Regulatory Commission to deprive people of life without due process of law, and if the Supreme Court turns its head from realizing this, as it did in declaring the Price-Anderson Act to be Constitutional, where are the guarantees that far worse injustices and violations of human rights will not be carried through in the future?

Personal Responsibility

In the USA, we have already accepted the policy of experimentation on involuntary human subjects. Every year, we introduce new chemical compounds of *uncertain* toxicity into the workplace and the biosphere. In the mid-fifties -- when the toxity of low-dose *radiation* was still uncertain -- we were testing nuclear bombs in the atmosphere and launching the Atoms for Peace program.

It should have been clear to me, even then, that both atmospheric bomb-testing and nuclear power constituted experimentation on involuntary human subjects, indeed on all forms of life. Instead, I am on record in 1957 as *not* being worried yet about fallout, and still being optimistic about the benefits of nuclear power.

There is no way I can justify my failure to help sound an alarm over these activities many years sooner than I did.

I feel that at least several hundred scientists trained in the biomedical aspect of atomic energy -- myself definitely included -- are candidates for Nuremberg-type trials for crimes against humanity through our gross negligence and irresponsibility.

THE FENCE-JUMPERS '...

Now that we *know* the hazard of low-dose radiation, the crime is not experimentation -- it's *murder*. Perhaps the "fence-jumpers" at nuclear power sites are the Paul Reveres of today, as they use their bodies to try warning us against *accepting* a policy of premeditated murder.

But apparently public apathy has something to do with *numbers*. If the numbers who die from nuclear power are "not too large," particularly if they die quietly without sporting a little flag saying "I am a victim of nuclear power," there is apparently nothing to alert people to their imminent danger of losing freedom, justice, and the inalienable right to life.

What would people think about Congress setting up a new Commission to permit licenses to be issued for a totally novel industry which would kill 1,000 persons per year? Would the people be concerned? What about 10,000 persons per year? What about 100,000 persons per year?

It is obvious to any rational person that when you permit the first murder-license, you have opened the floodgates. I say this fully cognizant that, one day, society may choose to permit the murder of some number of people to achieve a so-called "benefit." But one had better think *long and hard* about how such permission becomes permitted, or the consequences could be the loss of all freedom and justice in our society.

Where and when should people begin to raise questions about the subversion of justice and natural rights by laws passed by fallible men, and declared "Constitutional" by fallible justices on the bench? It is my opinion that the Judge in the Sam Lovejoy case was having trouble with this question.

... AND THEIR TRIALS

Recently, as just about everyone knows, there have been a number of "civil disobedience" actions against nuclear power facilities. Fence-jumping. Occupations. Blockades. In all these cases, it is alleged that "criminal trespass" upon private property has occurred, and the "disobedients" are being brought to trial for their alleged crimes.

"Must the citizen ever for a moment, or in the least degree, resign his conscience to the legislator? Why has every man a conscience, then? I think that we should be men first, and subjects afterwards.

"It is not desirable to cultivate a respect for the law, so much as for the right. The only obligation which I have a right to assume is to do at any time what I think right . . . Law never made men a whit more just; and, by means of respect for it, even the well disposed are daily made the agents of injustice . . .

"The mass of men serve the state thus, not as men mainly, but as machines, with their bodies. . . In most cases there is no free exercise whatever of the judgment or of the moral sense; but they put themselves on a level with wood and earth and stones; and wooden men can perhaps be manufactured that will serve the purpose as well. Such command no more respect than men of straw or a lump of dirt. . . Yet such as these even are commonly esteemed good citizens.

"Others—as most legislators, politicians, lawyers, ministers, and office-holders—serve the state chiefly with their heads; and, as they rarely make any moral distinctions, they are as likely to serve the Devil, without *intending* it, as God."

Henry David Thoreau's
"Civil Disobedience," 1849, in
the magazine, THE DIAL

In almost all cases thus far, the judges have held that expert testimony concerning the hazard of nuclear power *cannot be presented to the jury*. Stated succinctly, the action of the judges means that one is not entitled to present to a jury of his peers the reasons for his action.

But a person accused of a crime *must* be allowed to explain, with the help of expert witnesses when necessary, in what way his behavior was true to a law *higher* than the statute. When a judge denies the jury's right to hear and then decide for *itself* whether a "higher law" was really applicable to the alleged crime, the judge is denying the right (and duty) of jury-members and all other members of society to consider the higher laws -- by which I mean the very principles of justice and human rights from which laws are meant to be derived.

When juries lose the right to decide where justice lies, when juries must confine themselves to statutes, then our legal system is morally bankrupt and restraints on tyranny are gone. In earlier times, juries could be defenses against the king's tyranny because -- while the king could still make any unjust laws he wanted -- juries did not have to follow them. Recent rulings which deny the relevance of a law higher than the statutes, set us back to the Nazi era, or even to medieval times -- except that the king has been replaced by a few hundred legislators whose "principles" are rated below those of most snake-oil salesmen by many Americans.

At Nuremberg, we said that individuals have "duties which transcend the national obligations of obedience imposed by the individual state." Now our own judges are denying it, by preventing fence-jumpers from showing juries how a higher law was involved in their behavior.

I am totally indignant when some people suggest that we have to have a major *war* before we can invoke the Nuremberg Principles! Those principles are simply a statement of the obvious: there is a higher law, a standard of justice and morality, which enables us to evaluate particular statutes and to guide our own behavior, and consequently the higher law has a greater claim on our allegiance than the lower laws. To assert that we have to have massive carnage and "victory" before the Nuremberg Principles apply, is an absurdity. The Nuremberg Principles are statements for all men and all times.

In some states there is a law which holds that a defense is valid if it claims that a violation of law was committed to prevent a *greater* harm than the harm caused by the law-breaking. Other states do not have this statute. But I do not think the existence or non-existence of the "competing harms" statute is of any relevance whatever. It is simply obscene that a judge in any state, with or without such a statute, can prevent a person from telling a jury of his peers why he committed an act for which he is being tried. That is simply unjust on the face of it, no matter what the legal doctrine holds.

LAW PROFESSION: UNAWARE? OR SOMETHING WORSE?

At first I thought it might be just my ignorance of jurisprudence which made me fail to understand why it must be reasonable and just for judges to prevent the juries from hearing how nuclear power violates human rights. It did not seem just to me, but I thought, perhaps, I simply did not understand.

Then I thought back on that Judge in the Lovejoy case. He realized that the issue was of some consequence. Moreover, I have talked with lawyers for many of the "civil disobedient" cases, and uniformly they tell me *they* are incensed by such action of judges. So if lawyers find fault with the action of the judges, if the Judge in the Lovejoy case was concerned lest he had erred in preventing the jury from hearing the evidence about nuclear power, then there must indeed be a real issue here even to those who know the ins and outs of the "law."

The issue seems to me so profound that I would think the entire legal profession would be up in arms about it -- that is, if the legal profession is concerned about *justice*. Perhaps it has not yet come to the profession's attention, in which case citizens should be asking every lawyer they know what he plans to do about this issue.

If it has come to lawyers' attention, and they are still doing nothing about this issue, one had better ask himself how much faith he wishes to place in the legal profession to protect freedom and justice. Maybe that is no concern of the legal profession.

PROTECTING *OUR* DISSIDENTS

The Palo Verde Nuclear Plant is going to commit random murder of citizens in Arizona—indeed its murders won't, in the long run, be limited to Arizona. That is sad. But it is less important, by far, in my opinion, than the implications of the current scene for the murder of natural rights and justice in the United States.

The experience of Nazi Germany has taught us that rulers can use legislative bodies to pass laws which will "justify" the crime of genocide, the greatest subversion of justice and human rights. Even a cursory examination of what is going on in the Soviet Union in such cases as that of Anatole Shcharansky shows that law is being used to violate justice there in a similar fashion.

Americans should indeed treasure the freedoms we have, for they are so rare in the world. What is so dangerous is *denial* of American rights (described in this paper) to some of our dissidents and to some of our juries.

While a path of great length separates the denial of rights now in the U.S. from denial of rights in Nazi Germany and Soviet Russia, the path is a direct one—a continuum—and its distance will be more and more easily traveled if we fail to stop transgressions against American rights at their earliest stages.

The reason is simple: by the time transgressions have become flagrant enough to disgust or even to threaten all decent people, the price of resistance has gone up exponentially.

As the cost of resistance escalates from mere ridicule (as an "alarmist" and "exaggerator") to loss of jobs, promotions, and grants, resistance dwindles to insignificance. By the time resistance would have cost decent Germans their lives, virtually no resistance was made.

Transgressions against the human rights of *our* dissidents and against the rights of *our* juries have a significance, *whether or not* we choose to recognize it.

Freedom from slavery (such as that of Nazi Germany or the Soviet Union) and the preservation of human rights and justice, are issues of an importance which transcends by far the murders to be caused by nuclear power. It has been said, properly, that eternal vigilance is the price of freedom. I think the current handling of civil disobedience cases in the courts should be a warning that citizens had better have a hard look at how far we are going down the path to the loss of freedom and all our natural (human) rights by a blind obedience of the courts to "laws" which violate freedom, human rights, and justice.

If citizens do not insist on the higher law which takes precedence over all man-made laws, there is really not much left. We understood that at Nuremberg. We told the world that we understood that. How have we forgotten so soon?

BIBLIOGRAPHY OF ANTI-NUCLEAR DIRECT ACTION GROUPS

Catfish Alliance
Box 6306
Dothan AL 36361
(205) 574-5770

Abalone Alliance
c/o People Against Nuclear Power
844 Market St.
San Francisco CA 94102

Connecticut Citizens' Action Group
130 Washington St.
Hartford CT 06106
(203) 527-7191

Eastern Federation of Nuclear Opponents
317 Pennsylvania Ave.
Washington DC 20003
(202) 547-6500

Potomac Alliance
1746 Swann St.
Washington DC 20036
(703) 548-3341

Conchshell Alliance
3005 Bird Ave.
Miami FL 33113
(305) 443-9836

Bailly Alliance
6105 N. Winthrop
Chicago IL 60660
(312) 764-5011

Paddlewheel Alliance
4-223 Read Center
Bloomington IN 47401
(812) 337-1406

Citizen's Energy Coalition
3620 N. Meridian St.
Indianapolis IN 46208

Oystershell Alliance
7700 Cohn
New Orleans LA 70113
(504) 861-1626

Boston Clamshell
2161 Massachusetts Ave.
Cambridge MA 02140
(617) 661-6204

Detroit Safe Energy Coalition
691 Steward, #B-1
Detroit MI 48202
(313) 872-4024

Great Plains Alliance
1104 Coats St.
Columbia MO 65201
(816) 753-5370

Sunflower Alliance
4311 Holmes
Kansas City MO
(816) 753-5370

Clamshell Alliance
62 Congress St.
Portsmouth NH 03801
(603) 436-5414

SEA Alliance
c/o New Jersey SANE
324 Bloomfield Ave.
Montclair NJ 07042
(201) 744-3263

Citizens Against Nuclear Threats
106 Girard S.E., Rm 1210
Albuquerque NM 87106
(505) 268-9557

UPSET
Rt. 1, Box 121
Richville NY 13681
(315) 355-2512

Long Island Safe Energy Coalition
Box 972
Smithtown NY 11787
(526) 979-7392

Westchester Peace Action Coalition
(WESPAC)
100 Mamaroneck Ave.
White Plains NY
(914) 682-0488 or 949-0088

Ohio-ans for Utility Reform
Box 10006
Columbus OH 43201
(614) 461-0136

Trojan Decommissioning Alliance
215 E. Ninth St.
Portland OR 97214
(503) 231-0014

Keystone Alliance
1006 S. 46 St.
Philadelphia PA 19143
(215) 387-5254

Susquehanna Alliance
RD 1
Stillwater PA 17878

Palmetto Alliance
18 Bluff Rd.
Columbia SC 29201
(803) 771-9999

Armadillo Coalition of Texas
4525 Bisbee
Fort Worth TX 76199

Crabshell Alliance
1114 34 Ave.
Seattle WA 98122
(206) 323-2880

Live Without Trident
1305 N.E. 45 St., # 210
Seattle WA 98105
(206) 632-8323

Northern Sun Alliance
c/o Paul Wiegner
22 S. Barstow
Eau Claire WI 54701

League Against Nuclear Dangers
c/o Naomi Jacobsen
Rt. 1
Rudolf WI 54475
(715) 344-6158

HOW TO PASS
A TRANSPORTATION BAN

While scientists continue to speculate about catapulting radioactive waste into outer space, rocketing it to the sun or burying it in the polar ice cap or salt deposits, people all over America are taking action against radioactive garbage. In Vermont over thirty towns have banned the transportation or disposal of high level radioactive wastes within their boundaries, while the city of New York has banned the transport of spent reactor fuel elements, isotopes of plutonium and highly enriched uranium. In upstate New York the citizens of Jerusalem passed first a resolution, then legislation against transporting of nuclear waste and spent fuel rods within the town's boundaries. And Citizens Concerned About Nuclear Waste, the grass roots group responsible for the ban, is working hard for a county wide ban.

By refusing to allow high level waste to be transported through or buried within our communities, we can make it clear to the nuclear industry that the waste problem has not been solved. If there are no towns, counties or states willing to accept the death cargo, if there are no highways or roads over which it can be transported, the utilities will be forced to continue expanding their spent fuel pools and eventually will run out of space. Hopefully then they will realize that it is an exercise in futility to continue producing dangerous waste which cannot be transported or stored safely.

The federal government is aware of the grass roots resistance to the transportation of high level radioactive waste, and in the Federal Register of January 31, 1980 published proposed guidelines which would seem to prohibit some bans while letting others stand.* These guidelines are "proposed" only. A lengthy process of public hearings is necessary and perhaps changes will emerge from these hearings. The important point here is that the government can only preempt a local ban after a lengthy court battle, and in our opinion the most likely outcome of a higher court ruling would not be an outright prohibition on local bans, but rather a complicated ruling which would uphold bans that do not "unduly interfere with interstate commerce."

While the Department of Transportation is working out its guidelines, it is very important that we continue making decisions about our towns, villages and counties which demonstrate to state and federal legislators that we know the dangers of transporting and storing high level radioactive waste. If citizens everywhere ban together to form a blockade it will be difficult, if not impossible, for the nuclear industry to ignore our wishes.

This section is divided into two parts. Part one describes the activities of grass roots opponents of waste transportation and offers step by step information on how to pass a transportation ban in your community. Part two describes how to develop an information packet on the transportation of radioactive waste which can be used to convince legislators and other influencial community members that a ban on waste transportation is needed. The statistical information, scenario of a transportation accident, copy of the Jerusalem ban and questions about waste transportation can be used as models for your information packet or xeroxed and included in your packet. All of the information in this section will be useful to grass roots groups interested in working for a transportation ban.

---

*A copy of these guidelines can be obtained in any law library by asking the librarian for the Federal Register for January 31, 1980, or by writing the Department of Transportation. (See part two of this section for the DOT's address)

PART I

HOW TO PASS A TRANSPORTATION BAN
FROM CRITICAL MASS JOURNAL

By Lee Thaldos

The following article from Critical Mass Journal *offers a step by step formula for passing transportation curbs in your community. This formula has been used successfully by grass roots organizers throughout the country to pass curbs and/or bans in dozens of towns and cities, and will undoubtedly be used again and again in the on-going effort to protect our communities from the dangers of high level radioactive waste.*

Each year, there are more than three million radioactive packages shipped via U.S. roads and rails. With the number of nuclear transport accidents on the rise, public concern is growing, and an increasing number of communities are adopting nuclear transport restrictions to protect public health.

"Hot" shipments probably pass through your community more often than you realize. If your city or town has not yet adopted nuclear transport safeguards, it's time you did something about it.

This article can help you organize an effective popular campaign to adopt transport safeguards in your community. The techniques described were used successfully in the 1977 Vermont Town Meeting campaign organized by the Vermont Public Interest Research Group, as well as by other groups around the country. A successful campaign depends on a thoughtful, well-organized group of hard-working people who can develop and implement effective strategies and tactics. It is *not* dependent on vast financial resources.

Investigate

The first step is to conduct basic research to determine whether or not shipments currently pass through your community--or if any are planned. What materials do the shipments contain? Who is the shipper and what is their safety record?

Research must be thorough and facts correct to gain credibility and raise the issue successfully. One mistaken fact or misstatement can blow your credibility for the remainder of the campaign. Your goal is to educate the public with well-researched, honest information.

In addition to combing through reports and documents, you should speak with people who have responsibility for nuclear shipments. Health officials, transportation personnel and police all should have some knowledge of radioactive shipments in your community. While few people, including officials responsible for public safety, are aware of the extent of current and proposed nuclear shipments, these same people may well support restrictions based on their own lack of preparation.

Publicize

The next step is to pick the forum or forums which will best air the issue. The goal is to find a receptive public official or agency that will investigate or otherwise publicize the issue and take action.

A local health board might be a good forum. It might also be a concerned highway, police, or fire official.

This is one issue in which allies are unpredictable. Speak with many people, screen them, and if they are on the right side of the issue, work closely with them. These officials can really make a difference. You should try to elicit their support as constructively as possible.

When the issue begins receiving publicity, and becomes "controversial," your campaign is succeeding. Any news coverage, whether anti- or pro- in nature, is helpful because the primary goal of the campaign is public education.

There are basically two ways to adopt nuclear transport restrictions in your community. You can either go to your county commissioners, city council, supervisors or selectmen and ask them to adopt restrictions through the legislative process, or you can propose a referendum or initiative.

While the legislated bans are attractive because there are only a handful of people to "lobby" (the council or board considering the measure), citizens still must be well-organized to demonstrate support for the legislation. Another potential drawback is that unless officials considering the bans have been canvassed and strongly support meaningful safeguards, delay, a softened measure, or defeat may result.

No matter which route you choose, a public campaign to educate and involve people is vital. When the question of restricting nuclear shipments has been put to the voters, they have overwhelmingly favored safeguards. For example, Vermonters in 38 of 41 towns passed transportation bans in 1977 town meetings.

There are a variety of transport restrictions that have been adopted around the country. Often, these restrictions are something less than an absolute ban on all nuclear shipments. For example, cities have required police escorts, elaborate notification procedures, route approvals, emergency response plans and other safety measures.

A total nuclear ban can run into many difficulties, not the least of which is the transport of medical shipments. These are the radiopharmaceuticals, such as cobalt and barium, which are used for chemotherapy and other medical uses.

While some of these materials contain relatively high levels of radioactivity, medical shipments can easily become an emotional issue and are best excluded from transport safeguards.

Hartford, Conn. recently defeated a transport ban, in part because the across-the-board ban impeded medical shipments. As a practical matter, a transport ban which includes medical shipments (or defense-related materials) will diminish the chances of success.

If you are in a pro-nuclear area, you may wish to press for restrictions on transportation of *all* harzardous materials, including liquefied natural gas, chlorine gas and radioactive materials. While this type of measure may be more feasible, it will not focus attention on the volatile nuclear issue.

Whatever restriction you decide to push for, the wording is very important. The less complicated and more direct the language of the proposal, the more likely it is to succeed. If someone doesn't understand the wording, they are not likely to vote for it. A typical referendum question might read: "Should the city prohibit transportation of nuclear materials and wastes from commercial nuclear power plants in and through the city?"

Referendum

If local legislators seem unlikely to enact firm regulations, choose the referendum route. The issue must be qualified for inclusion on the ballot. Each jurisdiction has its own election laws. Consult an attorney, the city or town clerk, the city or county attorney or the Secretary of State to determine election law requirements. Many states have initiative or referendum laws which set forth qualification requirements.

In many New England states, town meeting procedures also permit ballot items. For example, in Vermont, five percent of a town's registered voters must sign a petition to place a question on the town meeting agenda. Or, the town's selectmen may vote to place the item on the ballot. Even if town or city authorities are able to place a question before voters on their own initiative without a petition, it is still best to gather signatures because you, the petitioner, then control

the language of the question, and the petitioning educates the electorate. It is actually the first public stage of the campaign.

Besides, if petitioners cannot get the relatively small percentage of voters necessary to qualify the issue for inclusion on the ballot, the issue isn't ripe.

Organizing

Once you're on the ballot, the rest is simple--you organize. Volunteers must be recruited and trained. The most important thing volunteers can do is canvass. You divide up the community into sections, and assign a canvasser or a team of canvassers to cover each area. The purpose of the canvass is to identify supporters and to get them out by telephone calls on voting day (not unlike a political campaign).

Canvassers should be thoroughly familiar with the issue. They should be prepared to anticipate and thoughtfully answer common questions, defend all information in your literature, and should be as courteous (and brief) as possible with each voter visited.

Endorsements from community leaders, public officials and union, professional and other civic groups are crucial. Identify key opinion leaders early--before your campaign is in full swing. These people and groups must be approached by your most articulate spokespeople and sounded out. Again, the earlier this is done, the better the results will be.

In Vermont, a few doctors were concerned about the issue, and organized nearly 40 doctors and health care professionals to sign (and pay for) quarter page ads in key newspapers. Later, the state Democratic Party also supported the bans publicly. Each time someone new endorsed the bans, this received news coverage and others jumped on the bandwagon. By pacing endorsements, you can "peak" at the optimal moment.

A media campaign (radio and/or television ads) may help, but that depends on your circumstances. Some stations will sell time based on a group's non-profit status, regardless of the ad's content. In Vermont, VPIRG prepared radio spots (using volunteer voices and studio time) and began airing them five days before the town meeting votes. This took the utilities by surprise, and proved quite effective.

A brochure or flyer should be distributed during canvassing and at the polls. Literature should be as brief as possible and totally accurate. It should explain the transport issue only, and not be a pamphlet explaining nuclear power in general or any non-transportation aspect of it.

Do not expect voters to read long political diatribes. Check election laws about distribution of political materials in the vicinity of polls. There may be some restrictions which could result in embarassing and damaging election complaints.

Group representatives should be available to speak to any local group. Conservation groups, historical societies and civic organizations are always looking for speakers. Radio talk shows are easy to appear on. Issue forums featuring pro- and anti- speakers are also excellent vehicles. Don't be afraid to fully air the issue, or even sponsor an event in which opponents have an equal opportunity to present their side of the issue. Just publicize the forum, and make sure as many supporters as possible attend. Be sure everyone brings an undecided friend or acquaintance.

Some Tips

A campaign should be well-organized, tightly orchestrated, and multi-faceted--a variety of tactics should be utilized. This "blunderbuss" approach has one very important advantage--it involves the maximum number of people in the campaign. By using a variety of tactics, people who, for example, may be afraid to go door-to-door will have something else to do. This is the most important job of any organizer--finding ways to involve the maximum number of people.

Win or lose, you will have scored a major victory. You will have identified supporters, allies and volunteers, and will have built an organization which can be mobilized for the next campaign. You will have begun educating the public on nuclear power.

When Montanans tried unsuccessfully to adopt a nuclear safeguards initiative in 1976, instead of giving up, they used the experience gained and tried again in 1978. They won. You can too.

HOW TO PASS A TRANSPORTATION BAN
FROM GROUNDSWELL

The following information is from "Groundswell, A Resource Journal for Energy Activists," available from Nuclear Information Resource Service, 1536 16th St. N.W., Washington, D.C. 20036. NIRS acts as a clearing house for information on nuclear power and, between the hours of 1:00 and 5:00 p.m., provides a toll free nuclear information hotline. The hotline's number is: (800)424-2477.

"How To Pass A Transportation Ban" describes the steps taken by The Charleston Palmetto Alliance and others to secure passage of a transportation ban in Charleston, South Carolina. This information should be useful to people living in large towns or cities who are considering working for a municipal transportation ban.

In 1978 the number of towns asserting authority over radioactive shipments rose again when Charleston, S.C. passed strict regulations over the transportation of nuclear materials through the city. Charleston—scarcely a hotbed of anti-nuclear fervor—became concerned when the Nuclear Regulatory Commission showed interest in using the city's port as the regular point of entry for weekly shipments of foreign nuclear wastes. The foreign spent fuel, imported under the Atoms for Peace program, goes to the government's Savannah River reprocessing facility to be recycled into nuclear weapons and reactor fuel.

With the number of spent fuel and other waste shipments rising each year, residents of towns and cities around the country are looking into the possibility of passing ordinances regulating transportation of nuclear materials. The Charleston Palmetto Alliance, which played a major role in organizing the Charleston effort, offers some advice on how to pass a municipal transport ban.

1. Collect and publicize accurate information about the nature of shipments through your area. In Charleston, the issue was relatively clearcut. The NRC conveniently solved the problem of public awareness when its inspection team arrived to check out Charleston's port. But in communities where actual or potential shipments are not so well publicized, research and public education must precede any legislative effort to pass an ordinance. Finding out what is actually shipped through your area may be difficult, but you need these facts in order to convince legislators and the public of the need for regulation.

2. Find out which local authority has jurisdiction over the flow of traffic over municipal roadways. This will be the enforcing body for the ordinance: the office in charge of issuing and denying permits, if that is included in the ban, or the body to whom notification of shipments must be made. Usually the town holds the authority; but if it's the county, then the proposed ordinance must be routed through county government.

3. If it is not already common knowledge, find out who is on the local governing body--the City Council or its equivalent--and how that body works. You should know in advance what every step of the legislative process will involve.

4. Find a sponsor among members of the governing body. If your local legislative process requires the ordinance to be introduced by a Council member, then this step is obligatory. Otherwise, it is optional but definitely a wise move. Needless to say, the more political weight your sponsor pulls, the better off you will be.

5. Draft the ordinance. This requires careful thought. Most of the major municipal transport bans so far have been based on the 1976 New York City ordinance, but the fact that most towns are not New York City makes minor modifications necessary in each case. Charleston, for example, determined that the city's budget would not permit setting up or enlarging an office for the purpose of issuing and denying permits, as was done in New York City. Instead of adopting a permit system, Charleston's ordinance distinguishes between three categories of radioactive shipments: materials banned outright (plutonium, enriched uranium, actinides, and spent fuel in quantities over 20 curies); exempted materials (medical and defense-related shipments); and all other radioactive shipments, which require notification of the Chief of Police.

6. Get the ordinance into the hands of the person who will introduce it, and give a copy to the legal counsel for the City Council as well. The proposed ordinance will almost certainly be debated on the basis of its legality--the town's right to restrict the flow of traffic that may well be part of an interstate transport system. But other towns have done it, and the Department of Transportation's upholding of the New York ban stands as an important precedent. Furthermore, the NRC's new draft regulations on spent fuel indirectly affirm local rights to restrict nuclear shipments, by stating that the existence of local ordinances will affect the NRC's choice of routes. (See: NUREG 0561, "Physical Protection of Shipments of Irradiated Reactor Fuel," June 1979.)

7. Follow the proceedings of the legislative process. The ordiance may be referred to a committee, and hearings may be held. At this state, it is advisable to line up some expert witness to testify on your side. To begin with, you need someone--a member of the community--who understands the proposed ordinance from a technical point of view. He or she must be able to explain exactly what the ordinace will regulate and why it is necessary in your community. This person will need a fairly good background in nuclear issues, in order to be able to explain what the banned materials are, what a curie is, and so on.

In presenting your case, it's helpful to provide estimates of what the consequences of a radioactive transport accident in your town would be. These estimates can be based on calculations drawn from government studies; simply plug in population and population density figures appropriate to your area. A study useful for this purpose is NUREG 0194, "Calculations of the Radiological Consequences from Sabotage of Shipping Casks for Spent Fuel and High Level Waste." (Available from National Technical Information Service, 5285 Port Royal Rd., Springfield, VA. (703) 557-4600. $4.50) This is the study on which the new spent fuel regulations are based.

Finally, you need someone who can explain how the ordinance would be enforced in the town. This could be a member of the enforcing agency, or it could be any member of the community who can convey the idea that enforcement will be simple and unproblematic, and will not require special technical expertise or a large investment of time and labor.

8. After this, it's basically a matter of lobbying to get the ordinance through. In some places this may depend on public participation and support, and in others, it may be largely a matter of playing local politics. In any case, it is certainly effective to have influential members of local government, such as the mayor, on your side.

Some towns have provisions for public input into the legislative process, such as a segment of time reserved for community members to express their opinions during each Council meeting. If this is the case, marshal all the community support you can get to appear at these sessions.

For further information, contact the Charleston Palmetto Alliance, c/o Susan King Dunn, 37-B Charlotte St, Charleston, S.C. 29403.

TRANSPORT RESEARCH: WHERE TO START

By Fred Millar

**No One Seems to Have Reliable
Figures on Nuclear Shipments**

To begin your nuclear transportation ban campaign you will need vital information. In this article Critical Mass Journal *suggests places to write for information about the transportation of radioactive wastes.*

The beginning of any nuclear transportation campaign is research. It is difficult, often tedious, but vital. The information you obtain can be used to make your case in the media, before your local and state representatives, and most important, before the citizens of your community.

Where to Start

The federal government has only general estimates of nationwide nuclear shipments. The Department of Energy, and the Department of Transportation (Federal Aviation Administration, Federal Highway Administration, Federal Railroad Administration, National Transportation Safety Board, Coast Guard, etc.) rely on a 1975 survey by Batelle Pacific Northwest Laboratories.

But that does not mean calling Washington will prove futile. When the Potomac Alliance contacted William Brobst, chief of DOE's Transport Branch, he said 10 percent of America's atomic shipments passed through Maryland, Virginia and the District of Columbia. Using a conservative estimate of 2.5 million shipments nationwide, that meant 250,000 nuclear shipments come through the greater Washington area every year.

Federal officials and agencies vary widely in their willingness and ability to help citizens with nuclear transport questions.

Info From States

State officials can provide valuable information, particularly if the state has laws requiring prior notification of nuclear shipments. State radiological protection bureaus, health departments, and police may have figures on shipments. Local health and police departments may also provide information.

Utilities' Final Safety Analysis Reports and Nuclear Regulatory Commission's Final Environmental Impact Statements can include transportation information. These can be obtained from NRC regional offices, or at the Public Document Room at the library nearest a nuclear plant.

Other leads on atomic transport are in the "Nuclear America" map produced by the War Resisters League and the "Nuclear Fuel Materials Movement by Highways" chart (1975) from U.S. Geological Survey.

Dr. Anthony Tse at the NRC's Office of Standards Development, (202-443-5946) has been very helpful in providing region-specific data based on the 1975 Batelle study. Letters to nuclear shippers may elicit some useful data.

DOE's Public Information Office will furnish some information on shipments of spent naval reactor fuel and other DOE wastes, as well as on foreign spent fuel shipments coming into the U.S. (mostly at Portsmouth, Va.) for storage at Barnwell, S.C.

One Approach

The proven method for estimating the volumes of atomic transport in a region is to determine a rough estimate of what percentage of national shipments come through a given area, and then to calculate the regional amounts for each type of nuclear material based on the national statistics for 1975-1978 and projections for 1980-2020.

Rough as these estimates are, they are the best anyone, including the federal government, can come up with today. Washington plans a better data collection system by 1985.

The difficulty with this technique will be getting regional estimates. An informed inquiry to DOE may be useful, but various citizen groups may have to enlist the help of local or congressional officials to persuade DOE to do regional calculations. Even so, some groups may have to do their own calculations. It is crucial for groups to be relatively conservative in both assumptions and conclusions, to maximize credibility with the press and local officals.

Applying Pressure

You may find that even "conservative" estimates of nuclear shipments will show huge amounts of radioactive material flowing through your community. With the potential health threat documented, you can move on to a public education campaign and start to put pressure on officials to protect the community from the dangers posed by nuclear transportation.

COMMUNITIES BLOCK SHIPMENTS

By William Reynolds

**Towns, Cities and States Protect
Themselves From Deadly Cargo**

The following article describes some of the things citizens are doing to oppose the transportation of radioactive wastes through their communities. This article is from Critical Mass Journal, *a monthly publication devoted to energy issues. A subscription to* Critical Mass *is six dollars a year and can be obtained by writing: P.O. Box 1538, Washington, D.C. 20003. Back issues of* CMJ *can also be obtained.*

Faced with inadequate federal regulations and mounting prospects for disaster, local governments across the country are taking the initiative to regulate, restrict or ban the transportation of dangerous radioactive materials.

Federal regulations have several shortcomings: they don't require notification to local or state government agencies of radioactive shipments, don't require emergency response plans for transportation routes, and provide inadequate monitoring of nuclear shipments.

Gaps in federal regulation on atomic transport are particularly dangerous because of the volume and toxicity of the shipments.

In 1975, 2.5 million packages of commercial radioactive materials were shipped in the United States.

The Nuclear Regulatory Commission estimates that by 1985 the number of radioactive packages shipped will jump to about 5.5 million.

Shipments of nuclear fuel cycle materials will more than quadruple from the current level of about 200,000 packages per year to more than 830,000 packages per year in 1985, if current trends continue. Radiopharmaceutical shipments will increase by only about 10 percent per year.

In the City

Many communities have already moved to curb radioactive shipments. For example, before 1976, shipments of spent nuclear fuel and large quantities of plutonium oxide passed through New York City.

Recognizing the potential public health threat posed by these lethal radioactive materials, the New York City Department of Health enacted regulations severely restricting large shipments of radioactive materials in 1976.

The regulation was designed so that it does not impede the shipment of radiopharmaceuticals and other radioactive materials needed for medical, research and industrial purposes.

The day the regulation went into effect a legal battle over its validity started. The Department of Transportation ruled in April, 1978 that due to a federal regulatory vacuum on

routing requirements for radioactive materials, the New York City regulation is legal.

A Trend

Now, citizens from Greenville, S.C. to Lane County, Ore. are carrying out successful campaigns to educate their communities to the hazards presented by the transportation of radioactive materials.

Local and state governments are increasingly receptive to citizen proposals to regulate nuclear shipments. Since the enactment of the New York City regulation, more than 50 communities and states have passed similar legislation.*

In northeastern Ohio, Shaker Heights passed an ordinance modeled after the New York City regulation. The city mailed copies of it to neighboring communities and within a few months, 15 had adopted the ordinance.

Thirty-eight Vermont communities enacted restrictions on town meeting day in 1977.

Louisiana prohibited the shipment of high level wastes into the state. Minnesota requires legislative approval before any nuclear wastes can be transported into the state for storage.

Legislation in North Carolina requires advance notification of spent nuclear fuel shipments entering the state. Notification is made to the Highway Patrol, which then alerts appropriate areas as to the time of travel, origin, destination, route and transport vehicle description.

Connecticut and Rhode Island both have legislation requring advance notification of all large quantity shipments, shipments originating from nuclear power reactors, and shipments which are required by DOT regulations to bear placards on the carrying vehicle.

Legislation requiring simple notification is valuable for two reasons: 1) It can alert emergency response teams and other government agencies to prepare for nuclear shipments and 2) It aids in future planning by providing information

---

*There are now more than 80 communities with transportation bans on radioactive waste.

on routes commonly used to haul nuclear materials, the types of materials which are carried and the numbers of shipments.

Across-the-board transportation limits also work to insulate communities from atomic shipments.

The Georgia State Transportation Board banned all tractor-trailer trucks from interstate highways through Atlanta, rerouting them instead around the city on the perimeter highway, I-285. As a result, truck shipments of nuclear cargoes are routed away from the densely populated inner city.

Dallas, Texas restricts the transportation of all hazardous materials, including nuclear materials, to specified routes around the city.

Washington Watches

DOT is not ignoring the local actions to regulate radioactive materials. In August, 1978 the department announced it was starting a rule-making procedure concerning highway routing of radioactive materials.

DOT has the authority to enact regulations which would pre-empt all other regulations inconsistent with federal regulations. If this happens, communities would lose the protection offered by existing ordinances which provide strict, local regulation.

But DOT could also recognize the authority of states and communities to regulate nuclear shipments. The department could enact stringent regulations to strengthen state and local laws.

A public hearing was held in Washington, D.C. in late November to solicit comments on the rule-making changes. Representatives from public interest groups, public health departments, fire departments, and bridge, tunnel and turnpike authorities outlined clear positions in favor of local authority to regulate nuclear shipments, and supporting the need for strict federal regulation.

Concerns centered around two issues:
1) The need for advance notification of nuclear shipments to emergency response personnel and 2) The need for particularly hazardous materials such as spent fuel rods to avoid densely populated areas, hazardous areas and hazardous industrial zones.

Nuclear industry representatives voiced support for the pre-emption of local ordances by DOT, claiming that existing federal regulations are sufficient to protect the public; and that local ordances only serve to impede interstate commerce.

Local routing restrictions which limit the use of bridges and tunnels, establish maximum weight limits, and which prescribe special routes have been in effect for years.

One area of agreement was that radioactive materials should be subject to restrictions according to their level of potential hazard. Most radiopharmaceuticals and other small shipments present little risk, and should not be subject to as strict routing regulations as more hazardous materials such as spent fuel rods and high level wastes.

Draft regulations should be issued by DOT for public comment later this year. Final regulations can be expected 10 to 12 months later.

In the meantime, local governments encouraged by active citizens will continue to enact ordinaces regulating radioactive shipments; a clear sign to DOT that people want strict regulation of radioactive materials shipments.

PART II

A transportation packet is important for alerting community leaders to the dangers of radioactive waste transportation. Because millions of Americans do not live close to a nuclear reactor, they feel relatively safe from the deadly effects of radioactivity. What they don't realize is that high and low level waste is being transported over the highways and railways and that an accident involving high level radioactive waste could destroy their community and possibly their entire family. Whether grass roots organizers decide to sponsor an initiative or referendum or work through the legislative process, a packet of information on the perils of waste transportation should be developed and sent or hand delivered to community leaders.

The following information can be used as background material on the problems of transporting radioactive waste, and included in grass roots packets. I would encourage people interested in a ban to study this material and to use at least some or even all of it for your information packet.

HOW TO DEVELOP AND USE AN INFORMATION PACKET ON THE DANGERS OF TRANSPORTING RADIOACTIVE WASTE

By The Trumansburg Rural Alliance

In our efforts to get transportation bans passed within the communities in which we live and work the Trumansburg Rural Alliance has developed a few ideas on how to pass bans which we would like to share with other grass roots groups. Although the situation may differ somewhat in each community, we feel that the advice we have to offer can be very useful to grass roots groups interested in passing transportation bans.

The Trumansburg Rural Alliance's interest in passing a transportation ban in the town and village of Trumansburg (the town *includes* the village) began after we saw an article in a local newspaper about a grass roots group in Jerusalem, New York. The article described how Citizens Concerned About Nuclear Waste had lobbied local legislators and rallied hometown support to pass a ban against the transportation of fuel rods or radioactive waste within the boundaries of Jerusalem. After discussing the article and deciding that one of our priorities should be working for a ban in our town and village, and Tompkins County, the county in which most of our members live, we wrote CCANW for copies of their newsletter and later we invited Yve Zinaman, the editor, to one of our weekly meetings. Yve arrived with maps showing the routes over which trucks carrying radioactive waste travel, copies of the Jerusalem ban, and many valuable suggestions on how to begin working on a ban in our area.

Inspired by Yve and CCANW's success, we began research into our area. We wrote the Department of Transportation and the Nuclear Regulatory Commission requesting information about waste transportation through our town and county. Specifically we asked whether trucks carrying radioactive waste were traveling either through our village on route 96 or our town on 89. We did not expect nor did we receive a prompt response to our request.

Because Citizens Concerned About Nuclear Waste had not proven that radioactive wastes were being trucked through their area *before* they began working on a ban, we decided to follow their example and begin work on our infomation packet. We first looked into the transportation routes which will be used if West Valley, New York, formerly a reprocessing plant, is reopened as a Temporary Away From Reactor Storage Site. We gathered statistics on the number of highway accidents involving radioactive wastes over a period of years, discovered that over 80 communities in the U.S. have already passed transportation bans, and began work on a cover letter describing why we wanted to work for a transportation ban in our community.

Our packet includes a cover letter, the scenario of a transportation accident by Marvin Resnikoff, an article from the *Philadelphia Bulletin* on radioactive waste spillage, a copy of the questionnaire used by CCANW, the Trumansburg Rural Alliance's "Questions and Answers About Transportation Bans," maps showing routes to and from West Valley, and a copy of the Jerusalem ordinance. We also included excerpts from a report on the Salinas Salt Basin showing that our area is being considered for waste storage.

After we assembled our packet we started looking for sympathetic legislators, clergymen, civil defense personnel and members of volunteer fire departments. We attended county, village and town board meetings and presented our case for a transportation ban. To our surprise we found many people anxious to receive our packet and willing to support a ban on the transportation of high level radioactive waste.

STEPS FOR DEVELOPING AN INFORMATION PACKET ON THE TRANSPORTATION OF RADIO-ACTIVE WASTE

1. Find out if a community nearby has passed or is considering passing a transportation ban.

2. Invite someone from that group to speak about passing a ban.

3. Attend village, town or county board meetings to see how they are run. If possible attend board meetings with the grass roots group in your area that is lobbying for a ban.

4. Write the Department of Transportaion, Nuclear Regulatory Commission, State Officials, Private Companies, for information about radioactive waste transportation in your area.

5. Collect articles and statistical information about the dangers of waste transportation. (Use the information in this chapter.)

6. Develop a cover letter for your packet in which you describe why you are concerned about waste transportation and what you want to do about it.

7. If another grass roots group in your area has already developed an information packet you may be able to use some or all of their information. For example, Marvin Resnikoff's scenario of a transportation accident would be as valid for grass roots work in California as it would in New York.

8. After you have secured newspaper articles, maps, statistics, a copy of an ordinance, written a cover letter and decided on what to include in your packet, have a stapling and collating party.

9. After mailing or hand delivering your packet to people who may be interested in working for a ban, wait a few days, perhaps even a week, and then call the recipients of your packet to determine their response. Invite them to your group's meetings, visit their offices, ask for their advice and comments on your packet and how to proceed with your lobbying effort.

10. Don't hesitate to revise the packet. Delete, add, rearrange, in short—edit. Then send it out again and again.

11. Meanwhile, begin working on an ordinance.

QUESTIONS AND ANSWERS ABOUT TRANSPORTATION BANS

By The Trumansburg Rural Alliance

Wherever the Trumansburg Rural Alliance has presented information on the dangers of transporting high level radioactive wastes, we have been asked certain questions. After much research into the various aspects of transportation bans, we feel confident to answer these questions. We also feel that a set of the following questions and answers would be a valuable part of a packet on transporting high level waste. Groups and individuals with questions about transportation bans are welcome to write the Trumansburg Rural Alliance, Box 50, Trumansburg, New York, 14886.

Q: How can a town, village or county pass a ban on radioactive material if this material is transported on state highways which run through or directly by the community?

A: There has been much controversy over this issue; however, our research has shown that unless your state has passed a statute which prohibits the passage of such a ban you *can* pass a ban. And even if there is such a statute you may be able to find a more recent ruling which would qualify or limit the statute and give your community grounds for passing a ban. If you live near a college or university you may be able to find law students who are interested in the transportation issue to help with the research into statutes.

Q: But would our ban be upheld in court if state or interstate highways are involved?

A: More than eighty communities throughout the United States have passed bans or restrictions on the transportation of high level radioactive waste. Thus far there has been no challenge to a ban on the issue of state or interstate roads. New York City's ban on the shipment by highway of large quantities of radioactive materials in or through the five boroughs of New York was challenged by Associated Universities, Inc., which operates Brookhaven National Laboratory at Upton, Long Island. AUI asked the Department of Transportation whether the ban was inconsistent with, and thus preempted by, the federal Hazardous Materials Transportation Act, which regulated highway carriers of substances like radioactive materials. On April 4, 1978 the Materials Transportation Bureau of DOT ruled to uphold the ban. Interstate and state highways pass through the city of New York.

Q. Who would enforce a transportation ban?

A: It is doubtful that your ban will be tested at this time because carriers of radioactive waste have stated they will not violate local bans. The Nuclear Regulatory Commission has also passed draft regulations on spent fuel which indirectly affirm local rights to restrict nuclear shipments by stating that the existence of local ordinances will affect the NRC's choice of routes.

Q. But isn't the Department of Transportation going to rule soon on just who has jurisdiction over the transporting of radioactive waste? If the DOT decides to give the states jurisdiction, is there any reason to pass a local ban?

A: Yes, there are several good reasons. The DOT regulations are already out in proposed form but even if they are passed it will be more than a year before they take effect. Meanwhile high level radioactive waste may be transported through your community and a ban would discourage carriers from taking routes through your community. Also, it is not clear what would happen if the DOT regulations are passed and your town village or county decided to enforce your ban regardless of the federal preemption. A test case would most likely be the result and the publicity from such a case would actually benefit the anti–nuclear movement. The more pressure local communities put on people and bureaus at the state level the better chance we have of keeping radioactive waste out of our towns and even our state. The state of Louisiana, for example, does not allow radioactive waste to be shipped or stored within its boundaries.

Q: If the ban is challenged in court, wouldn't the cost of such a challenge be too much for a small community to afford?

A: The cost of a radioactive accident would be much more. If a truck carrying high level radioactive waste was involved in a serious accident in your town the property damage could be hundreds of millions of dollars, and large numbers of people would become ill and eventually die of cancer. The cost of defending your ban is miniscule compared to the costs of a radioactive accident.

Q: What would a small town do with a truck carrying high level radioactive waste once the truck had been ordered to stop?

A: This is an important question because even when high level wastes are encased in casks 6 feet in diameter with 8 inches of lead shielding and weighing some 20 tons, radiation still leaks from the casks. It has been estimated that a person touching one of these casks for an hour would be exposed to the equivalent of seven chest x-rays. Serious consideration should be given to where the trucks will be stored and how people handling the trucks can avoid contamination.

Q: How does a community know when the waste will be traveling by or through that community?

A: There is no reassuring answer to this question. The carrier is required to notify state police when a shipment is being sent, but confirmation of the route a carrier will take is not confirmed until just before the carrier leaves a nuclear plant for a waste disposal site. The most common reason given for keeping routes secret until just before departure is the danger of sabotage or hijacking. Yet it would appear that one of the best ways to avoid sabotage or hijacking would be to notify local authorities along the route such as police, fire and civil defense personnel so they can make preparations to protect their community *and the waste.* If someone really wants to sabotage a truck load of radioactive waste, it would seem their chances would be improved if the truck were traveling with little or no local protection. The idea that keeping a route secret will prevent anyone from sabotaging or hijacking a truck carrying high level radioactive waste is poor logic. All this really does is keep small communities from knowing when they are in danger.

In order to keep informed about the latest developments at the national level on the issue of transporting high level radioactive waste, grass roots groups can call or write the Department of Transportation. The address and phone number of the DOT are:

For information on Department of Transportation Guidelines

> Marylyn E. Morris
> Regulation Specialist
> Standards Division
> Office of Hazardous Materials Regulation
> Room 8102
> 400 7th St. S. W.
> Washington, D.C. 20590
>
> Phone: (202) 426-2075

If you wish to make comments on DOT regulations or guidelines

> Dockets Branch
> Materials Transportation Bureau
> U.S. Department of Transportation
> Washington, D.C. 20590

COVER LETTER USED
BY TRUMANSBURG RURAL ALLIANCE

One of the most difficult things to do is write a good cover letter. This is the letter which the Trumansburg Rural Alliance sent to introduce our transportation packet. The details could be changed to fit your particular situation, but this can give you an idea of one possible format for your letter.

Dear Board Member,

The Trumansburg Rural Alliance has decided to join in the effort to establish a transportation ban for Tompkins County, New York. We base our decision on a number of facts. For example, if an accident were to occur in or near our area, 30 to 40 square miles of farmland, as well as numerous small villages, would be rendered uninhabitable for generations. Because no insurance company will insure homeowners against the event of a nuclear transportation accident, families forced to leave their land and homes would be left destitute, virtually refugees in their own country. To add to their misery many of these homeless people would have received a lethal dose of radioactive poisoning. Within a few weeks to a few years, they would die of cancer.

The situation in our area is made even more critical because state committees have recommended the re-opening of West Valley as a Temporary Away From Reactor Storage Site to help solve the problems created by the closing of sites elsewhere and the growing volume of high level radioactive waste around reactors. We also live near a prime target area for waste disposal, Connecticut Hill, and there is new evidence that high level radioactive waste will soon be routed through our region on its way from Chalk River, Canada, to Barnwell, South Carolina.

We hope you will read our packet carefully and join with the many concerned residents of Tompkins County who are working for a ban on the transportation of high level radioactive wastes. We would be very happy to answer any questions you may have regarding this very serious issue. Please feel free to call (607) 387-5182 or write the Trumansburg Rural Alliance, Box 50, Trumansburg, New York, 14886.

We look forward to hearing from you and to having you join us in preserving one of the world's most beautiful areas.

Thank you for your time.

The Trumansburg Rural Alliance

COVER LETTER USED BY CITIZENS CONCERNED ABOUT NUCLEAR WASTE

Here is another style of letter which can be used as an introduction to your transportation packet. This letter was used by Citizens Concerned About Nuclear Waste, the grass roots group responsible for passage of a ban in Jerusalem, New York.

Dear Legislator,

We have prepared this packet of information which we hope will be useful to you.

Our purpose in preparing it was not to argue the pros & cons of nuclear power but to find out what it would mean to have high level radioactive waste transported through our communities.

Included in this packet you will find:

----an illustration & description of the fuel cycle and at what stage the process becomes high-level.

----some questions we ourselves asked and found the answers to.

----a map showing how high-level waste has been transported in the past and may be again.

----West Valley fact sheet

----West Valley road map and aerial view of the current storage site

----list of accidents involving radioactive waste

----the analysis of a possible accident prepared by Dr. Marvin Resnikoff of Buffalo and referred to by Dr. Pohl at his meeting with the Yates County Legislature on August 21, 1979

We hope you will have the time to read all this information and will feel free to call us if you would want to discuss any of this information.

Sincerely,

Yve Zinaman, Secretary

NUCLEAR WASTE TRANSPORTATION ACCIDENTS ON THE RISE

By Richard Pollock

One Out of Three Mishaps Releases Radioactivity into the Environment

This article from Critical Mass Journal *gives statistics on the number of accidents involving radioactive waste, and shows that the incidence of such accidents is on the rise.*

In less than five years there have been more than 328 transportation accidents involving radioactive materials—with 118 mishaps releasing radioactivity into the environment, according to federal documents.

The number of radioactive shipments has been rising, as has their accident rate, the documents show. And while the mishap rate has dropped for air and rail shipments, highway accidents show a marked rise.

The volume of highway traffic in radioactive substances has been going up, and is expected to increase further as more nuclear power plants come on line.

Highway accidents accounted for 87 percent of all radioactive shipment mishaps during the first eight months of 1978.

The data, released by the Department of Transportation to the Critical Mass Energy Project under the Freedom of Information Act revealed that more than $115,000 in property damage was caused by the nuclear transport accidents. That figure does not include costs for rescue operations or decontamination.

A National Transportation Safety Board official told *CMJ* that the data reflects improvements airlines and rail companies have made to protect the public against radiological spills.

"I'm afraid," the NTSB official said, "that the same cannot be said of highway carriers."

The Worst Offender

The worst offender among all carriers of radioactive materials appears to be the Tri-State Motor Transit Company of Joplin, Mo. The firm reported to DOT that its drivers and handlers were involved in 152 mishaps with nuclear materials since 1974.

The second most accident-ridden carrier is Trans-World Airlines, with 19 reported mishaps. All other carriers had less than 15 accidents over the same five year period.

Tri-State specializes in shipping hazardous and explosive substances. An NTSB source described the company as a "blue chip" carrier, one with an excellent reputation.

DOT, NTSB and the Interstate Commerce Commission could not show how Tri-State compares with other firms in accident frequency.

Of the $115,284 worth of property damage caused by nuclear transit accidents, $97,895 was attributed to Tri-State. The remaining $18,000 was spread among 67 other carriers.

The 328 reported incidents over the four-year, eight-month period (DOT has no figures for the last four months of 1978) show that approximately three radiation shipment accidents occur every two weeks in the United States.

More and More Mishaps

The trend for incidents is rising. In 1974, an average of 1.2 mishaps were reported each week. By 1977, this rose 50 percent to 1.8 mishaps per week. In 1978, it was up to 1.9.

The increasing number of nuclear-related shipments can only partially account for the heightened accident frequency, federal officials say.

While airline shipments of radioactive materials have been rising by 15 to 25 percent a year, air transport has been accounting for a declining share of nuclear-related accidents.

Air transit's share of radioactive shipment accidents plummeted from 59 percent in 1974 to 13 percent in 1978.

Highway accidents, meanwhile, rose from 39 percent of all transit mishaps in 1974 to 87 percent in 1978.

The improved airline accident rate is linked to safety reforms enacted by the air carriers, notes Ludwig Benner, a safety specialist with NTSB.

The data furnished CMEP were incident reports filed by carriers with the Department of Transportation. DOT officials could not say whether all accidents had been reported.

The Hazardous Materials Transportation Act of 1977 and the Explosives and Other Dangerous Articles Act both require carriers to report accidents.

LIST OF CARRIERS OF RADIOACTIVE WASTE AND THEIR ACCIDENT RECORDS

One possible way to find out how much radioactive waste is being transported through your community is to write directly to the companies that transport this waste. Below is a list of carriers with a tally of the accidents from 1974-1978 for which each company has been responsible. Write to these companies. Ask what highways they use, how often they travel particular areas, what safety precautions they practice, etc.

NUMBER OF ACCIDENTS ATTRIBUTED TO EACH CARRIER 1974-1978*

Company	Number of Accidents	Company	Number of Accidents
Tri-State	152	A & H Truck	2
TWA	19	Seaboard	1
Delta	14	Hennis	1
N.W. Airlines	12	Quantras Air	1
Pac Inter Mt. 1	8	No. Central Air	1
American Airlines	7	Thunderbird	1
United Airlines	7	Airborne Mess.	1
Eastern	6	Chem. Nuclear	1
Allegheny Airlines	5	New Engl. Nuc.	1
Roadway	4	Garrett	1
Consolidated Freight	4	Point	1
Continental Airlines	4	Holland Motor	1
Braniff	3	Time-D.C.	1
Flying Tiger	3	Air New Zealand	1
REAE	3	Smith Transfer	1
Shulman Air	3	ET & WNC	1
McLean Trade	3	Century Mercury	1
United Parcel	3	Scandanavian	1
Lee Way	3	Farrell	1
Piedmont	2	Pinto Trucking	1
Jones Motor Div	2	Precision Air	1
Yellow Freight	2	McCormach Highway	1
Southern Air	2	Richards Messenger	1
Pam Am Airlines	2	Exxon Nuc.	1
Federal E	2	Mitchell Bros. Truck	1
Drake Motor	2	Universal Transcontinental	1
Eazor E	2	Salt Creek	1
Overnight Trans.	2	European Continental	1
Puralator	2		

*Figures are from Radiation Materials Incident Reports submitted to the Department of Transportation from 1974 to 1978. It does not include accidents which may have gone unreported.

GRASS ROOTS QUESTIONNAIRE

By Citizens Concerned About Nuclear Waste

A questionnaire such as this one used by Citizens Concerned About Nuclear Waste can provide a good deal of information in a very direct and clear way to recipients of your packet.

Q. What Is The Difference Between High-Level And Low-Level Waste?
A. During the mining, milling, enrichment and fabrication process of the nuclear fuel cycle uranium remains low-level. It is not until it has been made into assemblies and used in a power plant that it becomes high-level.

Q. What Exactly Is The Nature Of The High-Level Waste Which Has Been (And Probably Will Be Again) Transported?
A. It is the fuel rods from the power plant that have become so radioactively hot that they can no longer do the job they were created to do.

Q. What Radioactive Elements Do These Assemblies Contain?
A. Each shipment of spent fuel would contain 2 million curies of radioactivity in various forms, such as:

KRYPTON-85   STRONTIUM-89   STRONTIUM-90
CESIUM-137   CESIUM-134   IODINE-131
RUTHENIUM-106

and depending on the type of accident, the following deadly radioactive actinides may be released:

PLUTONIUM-238   PLUTONIUM-239
PLUTONIUM-240, Am-241 Cm-242 & 244

Q. How Will These Fuel Rods Be Transported?
A. They are very hot so they must continue to be kept cool just as they have been in the reactor and at the storage pools at the reactor site. They will then be loaded onto a truck or railcar and shipped in specially designed casks.

Q. Does This Storage Have Anything To Do With New York's Salt Beds?
A. No. We are referring to "temporary" storage at West Valley. Discussions about our salt beds relates to "permanent" storage.

Q. Would This Law Include Transportation For Permanent Storage As Well?
A. Yes.

Q. Who Is Responsible For These Shipments?
A. Until recently the Nuclear Regulatory Commission was responsible for shipments only at the loading and delivery points. The Department of Transportation (DOT) was responsible for the type of packaging that is used. But while the radioactive substance was in transit not a single federal agency actually regulated the handling of nuclear shipments.

Q. What Do You Mean "Until Recently"?
A. The Nuclear Regulatory Commission (NRC) has recently decided that these shipments are indeed dangerous and has passed regulations which they think will make it safer.

Q. What Are These Regulations?
A. Shipments must avoid transport through populations of 100,000 or more. There must be two drivers with each shipment, one must be with the truck at all times. Also the drivers must carry guns and not stop except for fuel.

Q. Why Should We Have A Law Banning Transportation Of High-Level Waste And Spent Fuel?
A. Because under the provisions of the Atomic Energy Act of 1954 sole responsibility for radiological emergencies rests with the states, counties and local officials. We do not believe we are ready for such emergencies.

(The hospitals, ambulance, fire and police have had no training in dealing with this type of emergency).

Q. Can We Repeal The Law Once It Is Passed?
A. Yes.

Q. Have There Been Any Accidents Involving Radioactive Shipments?
A. Yes. The DOT has recorded 200 such accidents since 1970.

Q. Can The Radioactivity Which Is In These Casks Escape?
A. (see analysis of hypothetical accident in this packet—This is the analysis that Dr. Pohl referred to when he met with the Yates County Legislature).

SCENARIO OF A TRANSPORTATION ACCIDENT

By Marvin Resnikoff

The following scenario is by Dr. Marvin Resnikoff, scientist and internationally known anti-nuclear activist with the Sierra Club. The scenario appeared originally in the Sierra Club's Waste Paper *and has been included in more than one grass roots information packet on the transportation of radioactive materials to show just how catastrophic an accident involving radioactive wastes might be. Copies of the* Waste Paper *can be obtained by writing: Sierra Club Radioactive Waste Campaign, Box 64, Station G, Buffalo, New York.*

A semi pulls out of the Indian Point reactor at noon. It is a hot, sunny day with a mild wind. The trailer is an open rectangular frame made up of heavy metal beams with a radiation symbol on the side. Inside the lattice work is a metal cylinder about three feet in diameter and 18 feet long. The cylinder contains one spent fuel assembly approximately 13 feet long from the Indian Point reactor. The hot spent fuel assembly is immersed in water inside the cask. The gross weight of the semi is close to 35 tons.

The truck wheels onto highway U.S. 9, heading north out of Peekskill. The truck is on its way to West Valley, New York. Just outside of town, for some unknown reason, the truck swerves into a bridge abutment.

The collision with the unyielding abutment sprawls the truck across the highway. Both drivers are knocked unconscious. The truck catches fire. Traffic is stopped. Within 15 minutes the County Sheriff arrives. A tow truck, ambulance and volunteer fire company are called. Passersby stop and gawk at the flaming wreckage. The ambulance and tow truck arrive after 15 long minutes. The sheriff and tow truck operator, braving seering flames, wrench the still unconscious drivers from the wreckage. The ambulance takes away the drivers. More tow trucks and the State Police arrive. The State Police are the first on the scene to note the radiation symbol. They call the state radiation hot line asking for advice. Thirty-five minutes have passed since the start of the accident.

The State hot line has been looking for the truck since communication was broken off. Because of the new security rules, the truck was to be in constant communication with the dispatcher. Sirens blasting, the volunteer fire company arrives. Hoses douse the flames from the spilled diesel fuel. The two trucks have difficulty moving the semi. Several of the tires are flat, deflated by the heat from the fire. Two others are flat even though the fire never reached them. The truck had blow--out tires to prevent hijacking.

The State hot line tells the State Police to keep people away from the truck until a radiation survey is made of the area. It is a long wait. The truck sits. Traffic is tied up. The police and Sheriff move the curiosity-seekers about 100 yds. away. The highway is now wet with water from fire hoses. No one thought to call Con Ed. Helicopters are enroute from Washington. Finally, they arrive about three hours after the accident.

In white space suits and wearing masks, the Washington emergency assistance crew does a survey. The Geiger counters record deadly amounts of radiation 100 yards from the truck. Because a flood of water was released to put out the fire, a leak from the cask has gone unnoticed until now. Three hours have passed since the initiation of the accident. The cask is now hissing ominously. Steam from an unseated pressure relief valve is escaping. The steam contains cesium. (Cs) Apparently the crash shattered part of the cladding around the reactor fuel. This cladding had become brittle during reactor operation. Con Ed had shipped one of their hottest spent fuel assemblies, cooled only 150 days.

Following the radiation survey, the helicopter crew tell the police to evacuate immediately a one mile radius from the accident, including a local high school. The hissing sound increases. The radiation levels in the area of the cordoned--off truck increase. Only the emergency crew has protective clothing but it offers no protection against gamma rays coming from the cesium. The cask begins to deform. Because of high radiation levels, the fire trucks cannot get close enough to hose down the cask in an attempt to cool it off.

Ten hours into the accident, about 10% of the Cs has been released. An explosion takes place. The cask is ripped open near the top. The radiation levels go off scale. Massive amounts of Ruthenium, Strontium, Cesium, Cecium and alpha-emitting actinides are released. Persons one mile downwind have an intense burning sensation in the lungs. All die of cancer, most of lung cancer. Many die within days but a veritable epidemic of lung and bone cancer and leukemia occur in the following years. Eventually, a 31 mile sector will have to be evacuated including Yonkers and upper Manhattan.

Editors Note: This accident has not happened but our research indicates that it can.

ANALYSIS: *The Waste Paper* analysis of this accident follows. For several reasons, the NRC believes that this accident cannot happen. They are misguided.

Can the cask leak coolant? Each spent fuel shipping cask contains a pressure valve. Under a fire or collision, the valve may unseat. According to a reference on this subject, (Siting of Fuel Reprocessing Plants and Waste Management Facilities, ORNL 4451, p. 5-4), "If a cask that has been designed for a water coolant is involved in a fire, it is unlikely that the outer cask seal can be maintained. Generally, such a cask contains a pressure relief valve. Once this valve is actuated, it is extremely difficult to reseat; therefore we must postulate that all the coolant will be lost in a fire." *Editor's Note: Only casks cooled by water have been used to date.*

In an accident where the water leaks out of the pressure relief valve, the remaining water will steam up and the cladding becomes embrittled during reactor operation and can easily shatter.

The NRC has stated that the cask has shock absorbing fins which will absorb the energy of the collision. However, the cladding becomes embrittled during reactor operation and can easily shatter.

NRC accident analyses assume a low number of fatalities due to a loss of coolant accident. The NRC assumes that the noble gases, Xenon and Krypton, will be released, but not the more "serious" radionuclides such as Cesium (Cs). *The Waste Paper* analysis shows that some of the radio nuclides will not remain in the fuel pellets during reactor operation, but will be vaporized and located in the gap between the pellets and fuel cladding. The vaporization temperature of Cs is 1240 degrees Fahrenheit, easily attained in reactors which reach 3600 degrees Fahrenheit. If the cladding breaks, the water-soluble Cs mixes with the water and steam. We assume during this initial phase of the accident that 10% of the Cs will be released with the steam.

With 10% of the Cs released over a ten hour period, the dose at 100 yds from the cask is very large. *The Waste Paper* calculates a dose of 36 rads/min to the lung due entirely to this 10% Cs release. It is not possible to stay at 100 yds for longer than 10 to 15 minutes without receiving a large lung dose. In addition, the Cs deposits on the ground and clothes and radiates "groundshine." The dose closer than 100 yards is higher, of course. We estimate that the police and volunteer firemen will die in several days. In addition to a lung dose due to inhaled radioactivity the Cs will enter the blood stream and provide a whole body dose. Also, the passing Cs gamma-emitting cloud will provide a whole body dose.

The NRC has stated that the cask will withstand a crash into a wall at 80 mph. However, the cask involved in the test had no heat producer within the cask, therefore, and no testing of the pressure relief valve. This accident assumes that the pressure relief valve will be jarred loose due to the shock of the accident, or to the pressure within due to the 1/2 hr. fire. As the leakage continues, and radiation builds up to the point where persons cannot approach the cask, the major release due to the Zr reaction follows.

The cask accident is similar to the Three Mile Island (TMI) accident, where an explosion took place 10 hours into the accident (see June/July, *The Waste Paper*.) The loss of water causes the interior of the cask to heat up. There is no ventilation and circulation in the cask to retard the process. As the spent fuel cladding increases in temperature, an interaction between the zirconium alloy cladding and steam takes place at about 1700 degrees Fahrenheit. The Indian Point spent fuel assembly normally puts out 9.14 Kw of heat, while the zirconium water heat out put is 106 Kw, more than ten times hotter. The heat within the cask vaporizes the fuel pellets. Within seconds, an explosive mixture of hydrogen gas similar to the TMI bubble is formed.

The cask cannot withstand the pressure of the reaction and splits open near the lid, leading to a major release of radioactivity.

The short term lung dose is very high one mile from the accident, and this is primarily due to Ruthenium--106. *The Waste Paper* calculates a lung dose of 3720 rems, within the first two days following the accident. This dose is sufficient to kill everyone within 1 mile of lung cancer, if not killed by another type of cancer first. Much of the radioactive material will remain in the lung and continue to radiate those who do not die immediately. The NRC would say that only 5% of the Ru--106 will be volatilized because the Ru has a high volatilization temperature.

While this is true of the Ru volatilization temperature, Ru will be in the form of oxides RuO_2, and RuO_3 and RuO_4. The latter two oxides have a very low volatilization temperature, while RuO_2 will volatilize after the Zr reaction.

The Waste Paper calculates the marrow dose, over a 30 year period, due to inhaled radionuclides at 1 mile from the accident will be 1967 rems. With almost certainty, this will cause leukemia in every individual exposed. Within the first two days, the dose to the bone skeleton will be 40,700 rems. The 30 year dose to bone skeleton is *extremely high*—100 million rems—due to the actinides and Sr-90.

A large section of land will be contaminated by the accident. *The Waste Paper* has performed calculations due only to the deposition of Cs, and to the consequent groundshine (gamma radiation) only. We do not account for other radionuclides, contaminated produce and milk, and the resuspension of the radioactive material. Assuming stable weather and a wind speed of 2.5 mph (2 meters/sec), the Cs will deposit itself within 31 miles, in a narrow wedge shaped formation. The whole body dose will be 43.5 rems/y, or about 350 times background. The NRC regulations require that no person receive more than 0.5 rems per year. It would take 194 years before the Cs contamination decayed to safe levels; the land must be evacuated for this period of time.

COPY OF THE JERUSALEM BAN

On April 2, 1979, a tiny town in the finger lakes region of New York State passed a ban against the transportation of high level radiaoctive waste, including spent fuel rods, through the town of Jerusalem. The Jerusalem ban has become a model for many small communities interested in declaring their opposition to the transportation of high level wastes. The Trumansburg Rural Alliance has included this ban in all its packets on transportation issues and would encourage other grass roots groups to do the same.

LOCAL LAW NO. 1 OF THE YEAR 1979
Town of Jerusalem

A Local Law to Prohibit the Transportation of High Level Radioactive Waste including Spent Fuel Rods

At a regular meeting of the Town Board of the Town of Jerusalem, Yates County, New York, held at the Town Barn, Guyanoga Road, in the said Town of Jerusalem, on the 2nd day of April, 1979, at 7:00 P.M. there were

PRESENT: John Payne, Supervisor
Donald Frarey, Councilman
Loretta Hopkins, Councilwoman
Edward Culver, Councilman
Ralph Scofield, Councilman

ABSENT: None

Mr. John Payne offered the following local law and moved its adoption:

WHEREAS, *the said John Payne introduced said law which has been in its final form upon the desks or tables of the members of this board at least seven (7) calendar days exclusive of Sunday, prior to the date hereof; and*

WHEREAS, *notice of public hearing in regard to said proposed law was duly published in the March 21, 1979 edition of the* Chronicle Express, *a newspaper circulated in said Town of Jerusalem; and*

WHEREAS, *said public hearing was duly held in the Town Barn in said Town of Jerusalem on the 2nd day of April 1979.*

NOW THEREFORE, *be it enacted by the Town Board of Jerusalem as follows:*

ARTICLE I
Title
This law shall be known as "Local Law to Prohibit Transportation of High Level Radioactive Waste Including Spent Fuel Rods."

ARTICLE II
Purpose
The purpose of this local law is to prohibit the transportation of high level radioactive waste including spent fuel rods:
(a) Notwithstanding any law, order or regulation to the contrary, no high level radioactive waste, including spent fuel rods from nuclear reactors, shall be transported into the Town of Jerusalem, Yates County, New York.

ARTICLE III
Penalties
Whosoever violates the provision of this law shall be punished by a fine of $1,000 or imprisonment for six months or both, and vehicles or equipment used in connection with the violation shall be seized. The Town does this under Municipal Home Rule Law number 2, Subdivision 2.

ARTICLE IV
This local law shall take effect upon its filing in the Office of the Secretary of State.

For additional information or copies of the New York City ban on transport of radioactive waste, contact: Sierra Club Radioactive Waste Campaign, Box 64, Station G, Buffalo, New York

LIST OF COMMUNITIES THAT HAVE PASSED BANS AND RESTRICTIONS ON THE TRANSPORTATION OF HIGH LEVEL RADIOACTIVE WASTE

All over America communities, counties, cities and even states are refusing to allow radioactive waste to be transported through their boundaries without a fight. They are challenging the arrogance of the federal bureaucracy which continues to hand down rulings to local communities on what these communities can do about their own safety! More than eighty communities have passed transportation bans and many more will pass bans during the next year, regardless of federal rulings.

The following list of communities have passed either restrictions or outright bans on the transportation of radioactive waste within their boundaries.

Here is a list of states and municipalities that have enacted restrictions on the transportation of nuclear materials. Most restrictions differ. They can range from simple prior notification to outright bans. Individual communities with restrictions are listed at the end of each state.

This listing was compiled by *Critical Mass Journal* with the assistance of federal, state and local governments, utilities, and citizen activists working in the nuclear transport field.

California--Escort required on the Golden Gate Bridge.
Colorado--Banned from Eisenhower Tunnel on I-70.
Connecticut--Certain statewide restrictions. City of New London.
Delaware--Prior notification required on Delaware Memorial Bridge.
District of Columbia--Banned from downtown highways.
Florida--Statewide prior notification and route approval required. Port of Miami.
Illinois--Strategic quantities of highly-enriched uranium or plutonium banned from O'Hare International Airport.
Indiana--Permit required on I-80-90.
Louisiana--Spent fuel shipments are banned. Tunnel and ferry restrictions on other shipments.
Maine--Permit required on Maine Turnpike where shipments are limited to daylight hours.
Maryland--Bridge and tunnel limits.
Massachusetts--Routing and time of travel restrictions. All shipments banned from Plymouth.
Michigan--Bridge and tunnel limits.
Minnesota--Statewide prior notification. Tunnel restrictions.
Nebraska--Routing restrictions.
New Jersey--Statewide routing restrictions with permit requirements. Bridge limits. Prior notification of shipments with 20 or more curies. Carteret Borough.
New York--Statewide permit requirements and route restrictions. No shipments on Thruway. Prior notification for certain bridges. Rockland County. New York City (strict limits).

North Carolina--Prior notification required for spent fuel shipments.

Ohio--Many cities with restrictions. Beachwood, Berea, Brooklyn, Euclid, Fairview Park, Garfield Heights, Highland Heights, Lakewood, Lyndhurst, Maple Heights, Mayfield Village, Middleburg Heights, North Olmsted, North Royalton, Richmond Heights, Seven Hills, Shaker Heights, South Euclid, Strongsville, University Heights and Vermillion.

Oregon--Prior notification of shipments with 10,000 curies or more.

Pennsylvania--Routing limits and permit requirements.

Rhode Island--Statewide route and hours limits, permits and certification required.

Texas--Tunnel restrictions.

Vermont--Statewide timing limits and prior notification. Many towns have passed restrictions. Andover, Barnet, Benson, Berlin, Brattleboro, Brookline, Brunswick, Charlotte, Chittenden, Clarendon, Fairfield, Fairlee, Grand Isle, Guilford, Halifax, Nancock, Jamaica, Marlboro, Middleton Springs, Mt. Holly, Newbury, Norwich, Pawlet, Pittsford, Plainfield, Putney, Randolph, Readsboro, Rochester, Shrewsbury, Stannard, Thetford, Vershire, Westhaven, Westminster, Weston, Woodbury and Woodford.

Virginia--Bridge and tunnel restrictions. Louisa County.

West Virginia--Tunnel restrictions.

BIRCH ON ENERGY

The John Birch Society has launched a pro-nuclear campaign spearheaded by a group called Push Our Wonderful Energy Resources, or POWER. The drive is led by former New Hampshire Gov. Meldrim Thomson, who suffered a stunning ballot defeat last November in an election that centered on funding the controversial Seabrook nuclear power plant.

With Thanks

This section was developed with the help of William Reynolds and Fred Millar, two safe energy activists who work on nuclear transportation issues.

More information can be obtained from:

William Reynolds
Nuclear Transportation Project
American Friends Service
  Committee
P.O. Box 2234
High Point, NC 27261
(919) 882-0109

Fred Millar
Potomac Alliance
P.O. Box 9306
Washington, D.C. 20005

TRANSPORT BIBLIOGRAPHY

For recipients of your packet who wish to do a little research on their own, a bibliography should be included which shows the many reports, books, maps and charts which pertain to the transportation of radioactive waste.

GOVERNMENT PUBLICATIONS

Final Environmental Impact Statement on the Transportation of Radioactive Material by Air and Other Modes, Office of Standards Development, U.S. NRC, NUREG-0170, Dec. 1977. Available from National Technical Information Service, Springfield, VA 22161. $12.00 Vol. 1, $16.25 Vol. 2.

Interim Report: Transport of Radionuclides in Urban Environs: A Working Draft Assessment, Sandia Laboratories. SAND 77-1927, May 1978. Available upon request from N. Eisenberg, Transportation and Product Standards Branch, Office of Standards Development, U.S. NRC, Washington, DC 20555.

Federal Actions are Needed to Improve Safety and Security of Nuclear Materials Transportation, EMD-79-18, May 7, 1979. U.S. General Accounting Office, Document Handling Facility, PO Box 6015, Gaithersburg, MD 20760.

A Review of the Department of Transportation Regulations for Transportation of Radioactive Materials, Oct. 1977. Available upon request from U.S. Department of Transportation, Materials Transportation Bureau, Office of Hazardous Materials, Washington, DC 20590.

Transportation of Radioactive Materials in the U.S., NUREG-0073, May 1976. U.S. NRC, Office of Standards Development. Available upon request from Division of Technical Information and Document Control, U.S. NRC, Washington, DC 20555.

Legal, Institutional and Political Issues in the Transportation of Nuclear Materials at the Back End of the LWR Nuclear Fuel Cycle, PNL 2430, Sept. 1977. Available from NTIS, $12.50.

Physical Protection of Shipments of Irradiated Reactor Fuel--Interim Guidance, NUREG-0561, June 1979. Donald Kasun, U.S. NRC, Office of Nuclear Material Safety and Safeguards. Available from NTIS.

Regulatory and Other Responsibilities as Related to Transportation Accidents, NUREG-0179, U.S. NRC, Office of Standards Development, June 1977. Available from NTIS, $3.50.

Spent Fuel Storage Requirements--The Need for Away-From-Reactor Storage, DOE/ET-0075, Feb. 1979. U.S. Department of Energy, Division of Spent Fuel Storage and Transfer. Available from NTIS, $5.25.

Everything You Always Wanted To Know About Shipping High-Level Nuclear Wastes, DOE/ET-0003, U.S. Department of Energy, Division of Environmental Control Technology, Jan. 1978. Available from U.S. DOE, Office of Public Affairs, Washington DC 20585.

Summary Report of the State Surveillance Program on the Transportation of Radioactive Materials, NUREG-0393, Los Alamos Laboratories for the U.S. NRC. Available from NTIS, $4.50.

Transportation of Radioactive Materials by Rail, Aug. 1977. Available upon request from Section of Energy and Environment, Office of Proceedings, Interstate Commerce Commission, Washington, DC 20423.

Special Study of the Carriage of Radioactive Materials By Air, NRSB-AAS-72-4, Apr. 1973. Available from National Transportation Safety Board, Washington, DC 20591.

An Overview of Transportation in the Nuclear Fuel Cycle, NWL-2066, Battelle Pacific Northwest Laboratories for ERDA, May 1977. Available from NTIS, $5.00.

Transportation Accident Risks in the Nuclear Power Industry, 1975-2020, EPA520/3-75-023 Office of Radiation Programs, Environmental Protection Agency, Washington, DC 20460.

Exposure of Airport Workers to Radiation from Shipments of Radioactive Materials. NUREG-0154, Feb. 1976. NRC. Available from NTIS, $4.00.

Measurement of Radiation Exposure Received by Flight Attendants from Shipments of Radioactive Materials, NR-DES-0001. Office of Standards Development, NRC. Available from NTIS, $4.00.

Environmental Survey of Transportation of Radioactive Materials to and from Nuclear Power Plants, WASH-1238, Dec. 1972. AEC. Available from NTIS, $7.25.

Survey of Radioactive Material Shipments in the U.S.A., BNWL-1972, Battelle Pacific Northwest Laboratories, Richland, WA, for the NRC. Available from NTIS.

OTHER PUBLICATIONS AND RESOURCES

Transportation of Radioactive Materials: Part VI of Radioactive Materials in California, The Draft Report of the Secretary for Resources State Task Force on Nuclear Energy and Radioactive Materials, Oct. 1978. Available from the Resources Agency, Suite 1311, 1416 Ninth Street, Sacramento, CA 95814.

The Uranium Accident of September 27, 1977: The Case for Emergency Preparedness Plans and the Adequacy of Transportation Standards, Public Citizen's Critical Mass Energy Project, October 31, 1977. Available from Critical Mass, P.O. Box 1538, Washington, DC 20013.

Fallout on the Freeway: The Hazard of Transporting Radioactive Wastes in Michigan, Marion Anderson, PIRGIM, Jan. 18, 1974. Available from PIRGIM, 590 Hollister Building, Lansing, MI 48993. Check for price.

Radwaste on the Roadway, VPIRG, Feb. 1977. Available from VPIRG, 26 State Street, Montpelier, VT 05602. Check for price.

Nuclear Cargo in North Carolina: What are the Risks?, Elizabeth Ragland, NCPIRG. Available from NCPIRG, P.O. Box 2901, Durham, NC 27705. Check for price.

Atomic Cargoes in the MD, VA, DC Area, November 1978. Potomac Alliance, P.O. Box 9306, Washington, DC 20005. $1.00.

Transportation of Radioactive Materials in the Western States, Western Interstate Nuclear Board, Mar. 1974. Available from the WINB, 2500 Stapleton Plaza, 3333 Quebec, Denver, CO 80207. Check for price.

Transportation Accidents: How Probable?, William A. Brobst in Nuclear News, May, 1973, pp. 48-54.

CHARTS AND MAPS

Nuclear Fuel Materials Movements by Highways, Sheet Number 15-435 of the National Atlas Series, U.S. Geological Survey. Available from the U.S. Geological Survey, Reston, Va. 22092. $1.50.

Nuclear America, War Resisters' League. Available from WRL, 339 Lafayette Street, New York, NY 10012. 75 cents.

Man-Made Radiation Hazards in the U.S.A., Women's Strike for Peace, 120 Maryland Ave., NE, Washington, DC 20002. Enclose a stamped, self-addressed envelope.

Flow Chart of the Nuclear Fuel and Weapons Cycles, Fellowship of Reconciliation, Box 271, Nyack, NY 10960, or Nuclear Transportation Project, AFSC, P.O. Box 2234, High Point, NC 27261. Enclose a stamped, self-addressed envelope.

TAKE NOTE:

The Publications listed as "Available from National Technical Information Service (NTIS)" are often quite expensive. The high cost may discourage the development of a reference library by private individuals. Congressional representatives can order these publications, free of charge. Check to see if your representative will obtain a copy for you before ordering from NTIS.

FOR FURTHER INFORMATION, CONTACT:
Nuclear Transportation Project
American Friends Service Committee
P.O. Box 2234
High Point, NC 27261
(919) 882-0109

HOW TO START AN INITIATIVE PETITION

For Americans who want to see a new law passed or an old law rescinded, the initiative petition is a way to get directly involved in politics. Over 100 cities and 23 states allow for initiative petitions, and a small Washington based group, Initiative America, is working for passage of a constitutional amendment that would allow citizens to propose national initiatives. In states that do not provide for the initiative, coalitions of citizens' groups are working to add the initiative to their state's constitution. At least ten states are considering bills allowing for initiative petitions.

The initiative petition has been used by anti-nuclear activists to defeat proposed nuclear plants and deny the utilities the right to charge for nuclear plants **before they have even been built.** *In the future the initiative petition will be used more and more to express the American public's disillusionment with nuclear power.*

ADVICE ON STARTING INITIATIVE PETITIONS

By Ray Wolfe

The following is a brief overview of how the initiative petition works, and a step by step procedure for starting an initiative petition drive by Ray Wolfe, Professor of Biology at the University of Oregon and anti-nuclear activist for ten years with a Eugene-based grass roots group, The Eugene Future Power Committee.

The initiative process works this way: A group of citizens interested in seeing a proposal enacted into law writes, usually with the help of an attorney, an initiative. After the initiative is certified by a government attorney and, in some cases, made available to the public for examination and/or possible challenge, it is placed on a petition which the group then circulates through the community. If enough signatures are affixed to the petition before a deadline the initiative is attached to a ballot and voted on in a local, county, or state-wide election. In the hands of well-organized citizen groups the initiative has proved a formidable strategy for opposing nuclear power.

For example, in 1969 a number of Eugene, Oregon residents got together to express their concern over a nuclear plant which was planned for their area by the local municipal utility, the Eugene Water and Electric Board. On November 5,1968, 80 percent of Eugene's voters had approved a measure authorizing construction of a nuclear plant. But the concerned citizens group, calling themselves The Eugene Future Power Committee, refused to accept the vote as a mandate for nuclear power and began working to prevent the plant from being built. The strategy they chose for their opposition was the initiative petition. The EFPC's initiative asking for a four year delay in design and construction of the proposed plant was approved by 52% of the voters on May 26, 1979.

But of course it wasn't all that simple. Once the initiative was certified for the ballot, an educational campaign was begun to convince area residents to vote "yes." One member of the Eugene Future Power Committee wrote about the campaign:

"We printed educational flyers and bulletins which we sent out in mailings, made available at public meetings and passed out at supermarkets and shopping centers and handed out door-to-door. We used TV spot commercials and one longer TV commercial featuring a fish biologist who warned of the effects of thermal pollution on salmon. Our members wrote letters to the editors and we placed

our people on radio and TV talk shows. We sponsored three public speeches by national experts on nuclear power, one by the famed Dr. John Gofman. We set up a speakers' bureau within our group and accepted all invitations to present our side of the story."

Like most grass roots efforts, the initiative process requires enormous amounts of time and energy on the part of citizen groups. But ten years after the moratorium passed in Eugene, the EWEB has no plans for future nuclear plants and has turned to cogeneration from a pulp mill for an additional source of electricity. For a more complete description of The Eugene Future Power Committee's initiative petition drive, see *The Nuclear Dilemma* by Gene Bryerton (Ballantine Books).

A more recent initiative victory was in Montana where INITIATIVE 80 gives the voters the right to: "Approve any commercial nuclear power plant, uranium processing facility, waste processing facility or waste storage facility." Although a similar initiative was defeated in 1976, the sponsors of I-80 (a group called Nuclear Vote) continued working to gather grass roots support for their proposal. In spite of the nuclear industry's attempts to distort the intention or factual basis of I-80, Montana voters passed the measure by a two to one margin. The utilities spent $234,082 to the sponsors' $16,331, but Nuclear Vote won a tremendous victory.

One very interesting aspect of the initiative process is how hard the pro-nuke forces will work and how much money they will spend to discredit a citizens' movement. Nuclear Vote provides a rather interesting list of major donations from opponents of I-70, the defeated '76 Montana Initiative: Babcock & Wilcox, makers of the reactor that nearly melted down at Three Mile Island, $10,000. Westinghouse, $12,000, and General Electric, $12,000.

In spite of a united effort on the part of big business, government and local utilities to defeat the efforts of grass roots organizations, initiatives have been passed. And even if a petition does not pass, many people have been enlisted in the struggle, others have been given the opportunity to see just how far the utilities will go to silence opposition and, perhaps, most important, notice has been given to the nuclear industry that we are willing to fight long battles.

By ordering a temporary moratorium on operating licenses and construction permits for new plants, the Nuclear Regulatory Commission has conceded that nuclear power is not safe. The assumption is, however, that with certain modifications those plants now on line and those planned for the future will be made safe. This, of course, is nonsense. But while the NRC plays out its deception, we can be working on initiatives which will make certain that another plant will never go on line in the United States.

Steps for an Initiative Petition

Step 1: Find out if your state is one of the states that permits initiative petitions.

Step 2: Develop the initiative statement. Competent and experienced legal assistance is essential for composing the initiative statement.

It is important to keep the initiative statement as simple as possible, and to limit the number of conditions imposed. An excellent initiative statement may be weakened by including too many restrictions or other conditions. Many potential supporters may be lost because they reject just one of several conditions imposed by the initiative.

It is also important to be realistic in formulating the initiative statement. Initiative sponsors often feel very strongly about the issues involved, and it is difficult at times to resist the urge to settle matters quickly and forever. It is necessary to be pragmatic when it comes to the electorate. An initiative proposing a moratorium, for example, is far more likely to pass than one proposing an outright ban. An educational campaign is more likely to convince people to vote for an initiative than an emotional outburst against nuclear power.

In addition to strategic considerations, it is important that the initiative's aims, plausibility and impact be carefully thought out. It is possible to have the entire project defeated because of some simple oversight or through inadequate study and research. Unlike laws formulated in the legislature, initiative measures, once certified for circulation, are not amendable before the election vote. Once certified for signature gathering by the government (municipal in our case), the die is cast and it must be voted on, flaws and all. This has been a problem with about 75% of the initiative measures we have studied. For example, one proposed initiative would have banned all nuclear plants in the state without qualification. Unfortunately, two university reactors would have been prohibited. It is very important that the intent, plausibility and legality be carefully worked out *before* composing the precise statement and *before* seeking authorization to circulate petitions for signatures. It is a lot of work to obtain the required number of signatures (usually 10-15% of the number of ballots cast in a recent municipal, county, or state-wide election) so careful composition of the ballot statement is of the utmost importance.

Step 3: After writing the initiative proposal there is usually a filing procedure and some sort of approval by the government attorney. There may be a waiting period during which the title is available to the public for examination and possible challenge to determine whether it is misleading or otherwise inadequate. The opposition challenged our title in court but was unsuccessful.

Step 4: If the initiative survives any challenges, there is usually some sort of authorization to proceed to collect signatures. Usually the required number of signatures must be obtained before some deadline. Signatures must be those of appropriately registered voters (in Oregon), and these must be witnessed by the petition passer. It is better to collect the required number of signatures plus 25% more to assure that the minimum number will be valid. Validity of signatures is certified by an election officer in Oregon. There is probably some variation in signature requirements from state to state.

Step 5: Before the deadline the signed petitions are presented, with copies of the initiative attached, to the appropriate elections officer for certification. This is usually a good time for a press conference.

Step 6: After the elections officer certifies that the measure qualifies, the real work begins.

It would be helpful to consult with politically experienced people about various aspects of the initiative campaign from the very beginning. An initiative movement is an excellent educational process for young and old. Appoint a suitable publicity person or committee and make as much use out of the media as you can. Press conferences are useful at the kick-off for petition passing and after the required number of signatures are obtained.

Of course campaign tactics will differ. We challenged the Waterboard members to public debate. The local newspaper threatened to publish just our arguments when the opposition was reluctant to participate. Naturally, the opposition formed committees. Some of their actions were amusing and not very effective. For example, one committee held a kick-off news conference *but refused to answer questions from the press.* Obviously they knew little about the basic issues. . . they just wanted electricity to "power progress."

At first, Waterboard members were reluctant to debate. Instead, people from the Portland office of the Bonneville Power Administration were used. Later this became so embarassing that Waterboard members agreed to appear on panel discussions. On other occasions they enlisted nuclear engineers from nuclear plant corporations to give talks and sometimes to debate with us on the issues. We worked hard to maintain credibility and accuracy. Also, we frequently appeared at *their* press conferences with immediate response statements from *our* point of view. We organized door-to-door contact, distributing literature throughout the city.

One amusing incident occurred at the outset of our campaign. The university television station invited a panel with both sides of the issue. The opposition failed to show but asked the master of ceremonies to tell us to take it easy. This infuriated us. One of our members stated that four out of five members of the Waterboard had been appointed shortly *before* election so that they could run with the advantage of incumbency. This sort of thing can be devastating. After this, they generally had someone to represent them, though frequently not a resident of Eugene.

This sort of campaigning can be fun particularly if you have real political hanky-panky to deal with. It is important to keep a close watch on the opposition and to record their statements which you can use against them at a future date. It may sound a little paranoid, but with special interests continually working to oppose grass roots efforts,

one must be a bit wary.

In summary, get the facts and publicize them in the simplest possible language. Run a rational and aggressive campaign. Attempt to contact all segments of the voting public. Door-to-door may be the best way. Set up mechanisms to identify and include other supporters not in your campaign group. Work to keep up group enthusiasm and education. A policy of distributing frequent updates to keep the group informed is a fine supplement to regular meetings. Exploit the local press; frequently they need something to talk about. Monday is often a sparse news day. Set up ground rules of conduct in the group to assure accuracy and to establish group credibility.

NUCLEAR INITIATIVES

The process of putting nuclear power on the ballot is similar in all 22 states with the initiative process. Citizens circulate petitions with the text of a proposed law. If a specified number of registered voters sign the petition, the measure is put on the ballot, to be approved or rejected by the state's voters. The following map indicates the 22 states with the initiative process.

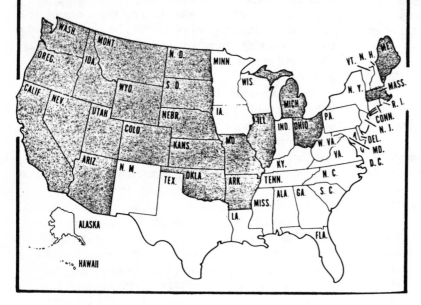

THE INITIATIVE TO CHANGE IN MONTANA

By David Schmidt

The following articles from Critical Mass Journal *will give you an idea of what citizens have been doing to pass initiatives in their respective states. Montana's I-80 has become an inspiration to anti-nuclear activists throughout the country and we can only hope that more initiatives like I-80 and Missouri's restrictions on Construction Work In Progress will soon be passed.*

In November, 1978, the citizens of Montana went to the polls and enacted the strongest state nuclear safety law in the country: Initiative 80. Now, after Three Mile Island, anti-nuclear activists in at least three states have taken to the streets with initiative petitions—the first step toward safe-energy lawmaking by popular vote in those 23 states where voter initiatives are allowed.

The initiative process is sometimes long, cumbersome, and frustrating if defeated, but it provides one of the most direct avenues for citizen action on nuclear power.

In Maine, Arizona, and Washington state, environmental groups decided months ago to follow Montana's lead and use the initiative process to pass anti-nuclear legislation. The people in Washington and Arizona patterned their initiative proposals after Montana, writing in strict safety requirements that have to be met before any new nuclear plants can be built. Example: Initiative 80 forces the utility company to take full financial responsibility for accidents. However, the Maine proposal is an out-right ban on nuclear power generation in that state.

The Montana Initiative will accomplish several important things: force utilities to agree to abandon limits on liability; compel plants to undergo extensive laboratory testing for safety defects; post a bond equal to one-third of the cost of the plant to cover decommissioning; put all certification decisions up for public vote.

In order for an initiative to be placed on the ballot, a certain percentage of the registered voters in the state (or municipality) must sign petitions asking for a public vote. Usually, the required number of signatures ranges between three and ten percent of the registered voters who voted in the last election. Petitioners have a limited time in which to collect the signatures. If the petition requirements are met, then the proposal must appear on the next election ballot; if approved by a majority of voters, it becomes law.

Industry propaganda

In Washington state, however, petitioners fell short by just several thousand signatures (out of a total of about 100,000 collected) when their deadline arrived in mid-July.

In Maine and Arizona, petition drives are now in full swing, with hundreds of activists circulating the petitions on street corners, in shopping centers, and anywhere else voters can be found. Prospects for success appear excellent in both states, despite industry propaganda campaigns aimed at stopping people from signing and circulating the petitions. Ray Shadis, leader of the petitioners' committee in Maine, claims that his opponents have resorted to spying on people to discourage them from becoming active in the petition drive. Since Maine's history is strongly footed in personal liberty and the direct democratic tradition of town meetings, Shadis believes this tactic may backfire.

The success of Montana's initiative and the prospects for a similar showing in Maine and Arizona may signal a rush of petition gatherings in other states. If so, one of the anti-nuclear movement's early citizen tactics could be reborn. In 1976, anti-nuclear initiatives appeared on the ballot in seven states, including California.

Only the measure in Missouri passed. It prohibited utilities from passing Construction Work In Progress (CWIP) costs along to ratepayers. The disappointing showing in 1976 led many to believe that the deck was stacked too heavily against the anti-nuclear forces for initiative campaigns to justify the effort. However, the public climate has been dramatically altered since Three Mile Island and petition-gathering campaigns may well face a more receptive electorate.

Wording is Crucial

There are 19 other states in which initiatives could be placed on the ballot for next year, if citizens there get petition drives started during the next few months.

Several factors should be considered before the campaign is started. First, after determining whether initiatives are legal in the state (or county or city), is the petition requirement. Wyoming, to take an extreme example, requires 15 percent of the voters to sign petitions. Experienced initiative petitioners say that it's almost impossible to get that many signatures in a state with its population spread out over such a vast area. Massachusetts, in contrast, requires only three percent and the voters are concentrated in a much smaller area.

Another important consideration is whether there are enough committed activists willing to work part or full time without pay, continuously, until the quota is reached—or, if such a group does not already exist, whether one can be organized.

A third consideration is whether activists can count on any political base within the state for help. For example, in the Montana campaign, influential Senator Mike Mansfield endorsed an earlier Montana initiative, contributing greatly to the final measure's success. Aligned with the need to draw support from political infrastructures is the campaign's ability to organize on a statewide level in order to reach substantial numbers of voters. Large states require well-coordinated networks of petition gatherers while smaller states still call for a highly organized campaign. Funding is a key question, as well; any campaign should be sure it can attract the dollars needed to take its best shot at success.

Finally, and perhaps most important, is the wording of the petition. Of paramount importance is that the measure be direct and simple so that citizens can understand what they are supporting and opponents cannot twist or exaggerate the initiative's intent. The failure of California's 1976 initiative has been partly blamed on a confusing text. The initiatives should be worded strongly enough to engender safe energy principles, but moderate enough to attract a broad-based coalition that can win in spite of the inevitable free-spending utility advertising campaign.

However, the rewards of a successful campaign can be immense— a virtual lock-out of the nuclear industry within a particular state.

Montana the Model

Montana voters approved Initiative 80 by almost two to one, refusing to be swayed by an industry media blitz that outspent the environmental groups 18 times over. Matt Jordan, a Montana activist who began organizing the Initiative 80 campaign right after a similar initiative was defeated in 1976, believes that Montana's bill should serve as a model for nuclear initiatives in other states. Copies of the initiative, plus background information, can be obtained by ordering the "Initiative 80 Press Kit" for 50 cents from the Nuclear Information Resource Service, 1536 16th Street, N.W., Washington, D.C. 20036. For information on how to organize an initiative drive, send a stamped, self-addressed envelope to Greater Washington Americans for Democratic Action, 1411 K Street, N.W., Suite 850, Washington, D.C. 20005.

David Schmidt is the executive director of Greater Washington Americans for Democratic Action.

CALIFORNIA'S KERN COUNTY VOTES DOWN NUKES 2-1

By Richard Pollock

"A rich, agricultural economy is synonymous with Kern County. Agriculture has made us strong . . . kept our economy healthy. That's why we must stop LA's land and water grab. We don't want LA's nuclear plant in Kern County and we can say so by voting NO on 3 at the polls March 7." – Radio Advertisement, Bakersfield, California.

BAKERSFIELD, Cal. — Voters in one of the nation's richest agricultural counties turned their backs on nuclear power March 7, 1978, piling up more than a 2 to 1 vote against the proposed San Joaquin nuclear plant, near Wasco.

An overwhelming 69.6 percent voted "no" to the special election ballot question: "Should a nuclear power plant be located near Wasco in Kern County?"

The vote, in effect, kills the proposed plant. Supporters of the measure garnered only 30.4 percent.

In 1976, the vote was almost exactly reversed, when Kern County residents helped defeat the anti-nuclear Nuclear Safeguards Act when it came up as a statewide ballot measure.

Political observers in California are predicting that the Kern County outcome will have major repercussions on the only other major nuclear plant proposal on the books in this state: the controversial Sundesert power plant near San Diego.

The March 7 vote is also expected to strengthen California Gov. Edmund G. (Jerry) Brown's anti-nuclear/pro-solar stance in the up-coming gubernatorial race. The national implications of Brown's energy stance have not been ignored in Washington.

On March 3, just four days before the Kern County vote, nationally syndicated columnist Tom Wicker described Brown as "the first national political figure to take the negative, openly and assertively, in the escalating national debate on nuclear power."

Brown himself vigorously defended his anti-nuclear stance while in Washington March 1 for the national governor's conference. He told reporters that his critics were saying "the same things they said about people who opposed the Vietnam War."

Historic Occasion

But the vote that should boost Brown's political stock in Washington took place 3500 miles from the nation's capital in one of the most conservative counties in California.

"This is an historic occasion for the people of Kern County," beamed David Bryant, one of the co-Chairpersons of the citizens' groups that opposed the power plant. "Our people have done their part in advising the Kern County Board of Supervisors that Kern citizens do not want LA's nuclear plant here in our county."

Bryant, a seasoned veteran of the county's fight for dependable water supplies, and Jack Pandol, the president of a sales agency for Kern County grape and fruitgrowers, have spent more than two years mapping out a strategy to defeat the four proposed 1150 Megawatt reactors. The Los Angeles Department of Water and Power, the plants' proponents, ruled out locating them in LA county, choosing instead to site them 100 miles to the north in rural Kern County. Kern's heavily irrigated plains have been called the world's most productive agricultural land.

Both Bryant and Pandol anticipated the plant's impact on water supplies, available agricultural land and jobs. Bryant, who also chairs the San Joaquin Agricultural Protection Council set out to document the consequences.

He calculated the project would remove 38,700 acres of agricultural land from production and divert 60,000 acre-feet of fresh water, dealing a $5.6 billion blow to the local economy over the 35 year life of the plant. They also found the county would lose 5,660 jobs and $68 million in wages each year.

In December, 1977 the Kern County Board of Supervisors, originally expected to approve the plant, took everyone by surprise and called for a public referendum on the reactors. They set the date of the special ballot measure for March 7 and indicated they would abide by the vote. The LA Department of Water and Power (LAWPD) also pledged they would not build the plant if the measure was defeated.

Anatomy of a Campaign

The campaign that ensued was a classic case of broad-based coalition building.

Opponents of the plant first sought organizational resolutions. The Bakersfield League of Women Voters, the county branch of the Cattlemen's Association, the San Joaquin Valley Supervisors Association, the Semitropic School District Trustees and the Kernville, Lake Isabella and Buttonwillow Chambers of Commerce passed resolutions opposing the plant.

Next, public officials were approached. Two of the five county supervisors announced that they were against the pro-

ject. The city councils of Shafter, Wasco and Delano passed resolutions opposing the San Joaquin reactor. The state Board of Food and Agriculture announced its opposition to the construction of power plants in agricultural areas. The Semitropic Cal. Water Storage District Board of Directors issued a resolution opposing the project so that water supplies would not be further depleted.

As the organizers secured resolution after resolution, their work began to pay off. The county's patriarchal figure, Republican State Senator Walter Stiern, surprised theLAWPD with a formal statement opposing the plant.

"Why risk land that can produce three crops a year," Stiern asked. "There are few areas that can do that. If you contaminate underground water supplies, you contaminate them for all the communities in the San Joaquin Valley."

The final blow came on February 21, when Governor Brown wrote Stiern that the "State of California is no longer an active participant in the WASCO Nuclear Power Plant."

Brown said, "California state law makes the proposed WASCO plant illegal, since there is presently no demonstrated way to permanently dispose of the dangerous radioactive waste such a plant would produce."

Bryant and Pandol organizationally established Kern Citizens for No on 3 to wage a precinct-by-precinct campaign. Between 500-600 county volunteers were mobilized to go door-to-door in an extensive canvassing effort.

Industry Bankroll

The supporters of the plant decided to counter this effort with full page ads and an expensive media campaign. The Los Angeles Department of Water and Power contracted Solem Associates of San Francisco, a public relations firm. The utility raised more than $165,000—four times the Kern Citizens' budget. Westinghouse, General Electric, Dow Chemical, Atlantic Richfield and General Atomic each contributed thousands of dollars.

Unmistakable Message

Pro-nuclear politicking and fundraising did not sway the voters. The final tally was 47,282 against the plant and 20,591 in favor.

"This victory has come after years of exhaustive studies, hearings and city-wide discussion," explains Bryant.

And, he added, savoring the success of the group, "there can be no question that our people made a clear and informed choice. The message is unmistakable: there is no need or desire for LA's nuclear plant in Kern County."

HOW TO START A LETTER WRITING CAMPAIGN AND DEVELOP A PHONE TREE

Many people can afford neither the time nor the money required to travel to Washington for a visit with their Congressman or woman. But most politicians do pay attention to the sentiments of their constituents. By reading hundreds and thousands of letters aides and secretaries to legislators can provide lawmakers with an idea of how people are feeling about a variety of political, economic and social issues. Therefore, it is extremely important for grass roots opponents of nuclear power to write their legislator(s). In this section Paula Curtin of the Trumansburg Rural Alliance offers a few suggestions to grass roots organizers on how to form a letter writing campaign. To form a letter writing campaign it is usually necessary to develop a phone tree, and in this section we also learn how to form a phone tree which will aid not only in letter writing campaigns but in other types of organizational work.

LETTER WRITING CAMPAIGN

By Paula Curtin

Writing letters to legislators is an inexpensive and remarkably effective way to influence public policy. In this section, Paula Curtin of the Trumansburg Rural Alliance offers suggestions for developing an effective letter writing campaign.

In order to change our local, state, and federal governments, citizens must voice their opinions loudly, and clearly to their legislators. Only by letting them know how strongly we feel and how many of us feel this way can we hope to have them represent us in their voting.

A letter writing campaign can be an effective tool. Bills have been introduced this way, while others have been amended, defeated and even passed, all due to public outcry in the form of letters, telegrams and phone calls.

How does a letter writing campaign work? First, make a telephone tree, that is, a list of your membership's phone numbers. If you do not belong to a group you can simply list people sympathetic to your cause, including acquaintances and strangers. Include the heads of other organizations because these key people will substantially increase the number of letters written by activating their own phone trees.

When an issue arises, contact each person. If the list is very long, get help from others who are willing to share the time and cost of phoning. Perhaps you could divide the list this way: You contact the first five names, a helper contacts the second five names, another the third five and so on down the list.

When the party answers the phone identify yourself, give a brief explanation of the issue about which you would like them to write and, when asked what should be said in the letter, have two or three key phrases prepared and give each person one phrase so that everyone does not write the same thing. For example, say either "Include Tsongas Amendment to S-668," or "Allow State Vote on Energy Authorization Bill." Then slowly recite the specific name, address, and zip code to be used. Give a reasonable deadline for writing if a deadline has not already been established. This ensures the arrival of correspondence in bulk for the greatest impact. If time does not permit the mailing of letters, recite the phone number of the legislator's office, or Western Union's Telegraph number. (Note: Political telegrams are cheaper than regular telegrams.)

Last, instruct the party to contact three other people, whose names do not appear on the list, such as friends, relatives, or co-workers. Have each of those three, contact three of their own. . . . . and so on. The further this branches out the greater the public outcry on Capitol Hill.

Let your voice be heard! Organize a telephone tree and letter writing campaign, now.

PHONE TREES

In the battle against nuclear power information is continually changing. New guidelines are being developed by the NRC and DOT regarding the various phases of the fuel cycle and these guidelines are released for public scrutiny, new legislation is proposed by members of Congress which grass roots activists will support or oppose, new defects are discovered in nuclear reactors, etc. For these reasons grass roots activists spend a lot of time on the phone. In this article taken from Critical Mass Journal *grass roots activists can discover a method which will make their phone calling a bit easier.*

A telephone tree is a simple way for us to communicate with one another and to organize ourselves so we can effectively mobilize for action. One such action may be required when a key congressional vote is imminent.

When a congressional vote is scheduled on an energy issue, it is desirable to get as many constituents as possible to quickly contact their elected officials by letter, telegram or telephone. It is not often possible for one concerned individual working alone to generate sufficient support in a short time. A telephone tree divides the responsibilities among concerned citizens and ensures that the people who need to know are aware of the safe energy position. In the process, no one should have to carry a burdensome load of work.

Many citizen groups have successfully utilized "Telephone Trees" to contact dozens of individuals and urge them to reach their congresspersons on upcoming votes. Very simply, when an Action Alert is sent out to the lead coordinators of the Telephone Trees, each coordinator contacts their key people who in turn contact other people. Through this system dozens of people are alerted and can begin sending letters, telegrams or phone calls to Congress or, in some cases, to a government agency or the White House.

How The Tree Works

A Telephone Tree is made up of a lead coordinator (the trunk), several branch coordinators (the branches), and callers (leaves). An effective tree should have one "lead coordinator," 5-7 "branches," and each branch should have 5-7 "leaves."

The tree operates when the lead coordinator receives an Action Alert, and phones the message to all of his or her branches. Each branch phones the message to each of his or her leaves. Then the lead coordinator, the branches, and the leaves ALL write, telegram or telephone the message to their elected officials.

1. *Before you begin:* Have on hand the name, address, and phone number of the representative in your group's district, the committee he or she is on and, if available, relevant facts from his or her past voting record. It is also helpful to have a copy of the Action Alert and Legislative Update available which can explain the issues behind the vote or action.

2. *Activating the tree:* Once a decision has been made to use the tree, the process is simple. The lead coordinator calls the branch coordinators. The branch coordinators call their list of leaves. The leaves call any people they have brought into the tree. And everyone sends the message to their representative.

3. *Write down the message:* It is important that all along the line people write the message down and read it back to the person calling them so that it gets communicated accurately. Always include in a statement the bill name and number, the congressperson(s) to be contacted and their addresses and phone numbers, and the request or statement to be made to the congressperson. Always suggest that the calls, letters, or telegrams be dispatched as soon as possible unless you have some specific timetable.

4. *Be persistent:* If lines are busy or people are out, try again. Branch coordinators should call back their contacts daily until they know that the messages have been sent.

5. *Spotcheck:* The lead coordinator should have a list of all the branch coordinators and their leaves. Lead coordinators should make a few calls to check to see if all jobs are being done effectively.

6. *Reverse flow of calls:* Everyone should be encouraged to call their coordinator if they have any questions about the status of the legislation, the ground rules of using the tree, etc. This way the branch coordinators can get feedback from the leaves and the lead coordinators from the branches.

7. *Keep the tree in gear:* The tree should be used at least once every 6 weeks.

8. *Give your people support:* Coordinators and branch coordinators should remind everyone that their call(s) can make the difference between winning and losing; they should be regularly informed about the bill's progress and of any change in their member's position.

Building A Tree

1. *Ready made trees:* If you already have a phone contact system for alerting your members about meetings or actions, consider using it to generate letters, telegrams or phone calls to your Congressperson in Washington about energy legislation. Make sure that everyone receives copies of the alert materials and understands their roles. You may want to explain the system at the next meeting.

2. *Create a tree from a group's membership:* Call a meeting or use your next scheduled meeting to talk about the telephone tree.

a. Pick out in advance the key people who will be your branch coordinators. Make sure they read and discuss this guide with you.

b. If your group is a federation of other groups, clubs or chapters, you may want to use the existing structures as the branches of the tree. Have the leader or executive committee member from those clubs or groups be the branch coordinators.

c. After you explain the tree concept, you can announce the names of the chosen branch coordinators. This will tend to create a snowball effect of those willing to take on the easier task of being leaves.

d. Pass around a sign-up sheet for those interested and announce a very short meeting at the end of the meeting to get the tree organized.

e. (Disregard this section if you wish to use existing club structures for your branches.) At the organizational meeting, divide leaves equally among the branch coordinators. Or if some coordinators have more time than others, you might consider weighting their branches with more leaves. In general, you do not want the branch coordinators to have more than seven calls to make unless they are willing and hard workers. If you are lucky enough to have more than seven per branch you might try to find another branch coordinator or two to reduce the average number of calls to the ideal 5 to 7.

f. Make sure you have a copy or list of the districts in case people in your group come from more than one district. If this happens, try to create separate trees for each district.

g. Give each person on the tree a copy of the tree plan. Have each person write down their branch coordinator and lead coordinator's name and telephone number.

h. Encourage the leaves to find one or two other people who might want to join the tree. Any people that they find should be given the branch coordinator's number to call so that they can receive copies of telephone tree materials. When the new people become part of the tree, it should be the responsibility of the person who brought them in to call them. This is a good way to build leadership while at the same time expanding the tree.

3. *Creating a tree without an existing group:* If you want to build a tree but do not have a large enough group to generate 30-50 correspondences, you can do several things. The easiest is to get five members of your small group to serve as branch coordinators. Then assign them the responsibility of getting 5-7 others to participate. Have a meeting with everyone to explain the tree, pass out materials, explain the issues, etc. A second way to literally pull a telephone tree out of the air is to go to the meeting of an existing group. Call church groups, senior citizen groups, consumer groups, block clubs in your area who are sure to be interested in energy legislation and ask if you may make a presentation at the next meeting. Bring Telephone Tree materials and follow the instruction in number 2 above. Try and get the leaders of the groups to promote the concept to their membership and suggest to you the people who would make good coordinators.

4. *Expanding the Tree:* You should be continually seeking to expand the leaves, branches and trunk as your groups' campaign progresses. Each time an individual expresses a good deal of interest in your work, ask if they would want to become part of the alert tree. You must make sure, however, that they will indeed respond to your Alerts. A tree of rather disinterested or only mildly interested people will cause an enormous amount of work and minimal results.

HOW TO SHOW THE CROSS COUNTRY TRAVELS OF RADIOACTIVE ISOTOPES

Although radiation is odorless and can not be seen by the naked eye, there is a way to demonstrate that it travels long distances from where it is first released. Activists in different parts of the country have used balloon releases to show that radioactivity does not simply disappear into the stratosphere but travels hundreds, even thousands, of miles. No one is safe from the deadly poisons which nuclear power plants routinely release into the air and water.

BALLOON RELEASES

By Fred Wilcox

Here are a few tips on how to use balloon releases to show that radioactive isotopes travel hundreds and thousands of miles and in patterns that are not always predictable.

In his recently published book, *IRREVY*, John Gofman warns that if the United States were to build and operate 1000 nuclear plants and if the containment perfection for these plants were 99.9 ("meaning a loss to the environment of only one-thousandth of the long-lived radioactivity.") we would "add roughly 198,000 extra fatal cancers annually in the U.S.A. alone for each year such plants operate."* "Lastly, it is known and documented that radiation-releases are *planned* and occur *regularly* at operating nuclear facilities, and that population exposure occurs as a result. . . This doesn't begin to include all the accidental spills and unmeasured releases of radioactivity. On every occasion where a measurement was made the industry was caught short."**

---

* *"IRREVY" An IRREVERENT, ILLUSTRATED VIEW OF NUCLEAR POWER,* John Gofman, Committee for Nuclear Responsiblity, 1979, p. 100.

** Ibid., p. 38.

Although we know that radiation is routinely released from nuclear plants it is still difficult to convince large numbers of people that nuclear power is dangerous. Unlike other carcinogenic pollutants, radioacitivity cannot be seen with the naked eye, felt by the human hand, or experienced as an unpleasant odor. No smog surrounds a nuclear plant, and the cooling towers do not belch black smoke—atomic power plants *appear* clean. And the nuclear industry has made the most of this deceptive appearance, declaring that not only are nuclear plants clean, but also so safe you could "eat off the floor of the control room."

A balloon release is one way to refute the old adage that "What you can't see won't hurt you." And it is a way of showing that even if you live in a community several hundred or even thousand miles from a reactor you are still in danger. There is no assurance that the wind will be blowing away from your community, and even if it is radiation travels long distances and in patterns that are not always predictable. *Someone's* community will be affected.

THIS BALLOON Symbolizing radioactive fallout was released near (name of facility) on (date). Fallout from a nuclear accident may travel this far. How far did this balloon travel? Please let us know by mailing back this card.

Date Found _____ Time _____
Balloon found at _____
_____

If you would like more information, please check this box ☐ and be sure to indicate your name and address on the other side.

Radiation (even in tiny amounts) has been shown to cause cancer.

Releases of small amounts of radioactive material from nuclear facilities is routine. Did you realize they came this far?

Up-up and away. Card bearing balloons take off from Seabrook, N.H. August 6, 1977.

Someone's cells will be damaged by the invisible poison from a nuclear reactor.

To demonstrate that radioactivity travels far from the point where it is released, grass roots groups in different parts of the country have sponsored balloon releases. Rocky Flats, Colorado, Montague, Massachusetts, and Fulton, Missouri, are just a few of the areas where up to a thousand helium filled balloons have been launched. Attached to each balloon was a card.

The cards are addressed to the sponsoring organization, with space for the name and address of the finder in the upper left hand corner.

The organization sponsoring a launch can hold a press conference before the release and/or, when a number of cards have been returned, write a press release. One week after their launch, Missourians For Safe Energy wrote:

'1000 Helium balloons, symbolizing radiation, were released from the construction site of the Callaway Nuclear Plants near Fulton, Missouri on Sunday, November 13, 1977, and have traveled to the bi-state St. Louis metropolitan area, (100 miles East of the plant), and have also landed in eight states between Missouri and the Atlantic Coast. The balloons, released by Missourians for Safe Energy (MSE—an environmental group concerned about both routine and possible accidental emissions from the plants), had cards attached asking the finder to inform the group as to where the balloons traveled.

Balloons landed in Illinois, Indiana, Kentucky, Ohio, West Virginia, Pennsylvania, Maryland, and Virginia. (See attached list of landings.) Just as radioactive material from nuclear weapons tests enters the upper atmosphere and then falls-out thousands of miles away, nuclear materials emitted from a reactor can, if weather conditions are right, travel extremely long distances.

The organizers of the balloon release were nonetheless quite surprised at how far their balloons traveled. Said Steve Johnson of Columbia, Mo., an MSE member: "It points out to the people of St. Louis and to people everywhere that we can't isolate ourselves from the environmental impact of nuclear power by isolating nuclear plants in sparsely populated areas." The balloons were released at 3:30 PM (CST) on Sunday, November 13th following a safe energy rally attended by approximately 100 people. Speakers at the rally called for the cancellation of the Callaway project and the development of solar and other renewable energy sources. Many respondents, apparently surprised at the distance and speed at which the balloons traveled, also included the hour of the day they found the balloon. One balloon, returned from Tappahannock, Virginia on an arm of the Chesapeake Bay, completed its eight hundred mile journey in less than 24 hours. It was found at 4:00 PM (EST) on Monday the 14th of November. In all, one thousand balloons were released. As of November 21, twenty seven cards have been returned from nine states." (The release then listed the places where the balloons were found.)

A balloon release is a relatively simple event to organize and, when the cards have been returned, provides grass roots groups with evidence to counter the argument that reactors built in sparsely populated areas are safe. As long as *one* reactor is on line, radioactivity will continue being released into the air we breath and the water we drink.

STEPS FOR A SUCCESSFUL BALLOON RELEASE

1. Decide on a site for a launch. If you live near a nuclear plant, enrichment facility, uranium mine or waste dump you could release your balloons nearby. If you do not live near a site, try to determine if your area is targeted for storage, mining, plant construction, reprocessing, or enrichment, and begin advertising the dangers to the public *before* the ground is ever broken.

2. After you have decided on the launch site, purchased balloons and printed the cards which will be attached to the balloons, invite the press to observe the release. Invite neighbors, friends, community leaders, utility employees, nuclear plant employees, hundreds of school children. Hand out flyers explaining how radioactive isotopes are carried by air and water to distant places.

3. When a number of cards have been returned to your organization, try to determine whether there were definite patterns of flight. To demonstrate how radioactivity travels it would be good to make a graph or maps showing where your balloons were released, the path or paths traveled, and where they were found. The graph or map could be sent to newspapers along with a press release.

4. Try to appear on your local television or radio programs. If there is an anti-nuke scientist in your area or expert in the effects of low level radiation, invite him or her to go with you. Take charts, maps, graphs, and all the scientific data you can on how radioactivity is getting into our water and air.

5. Encourage other grass roots groups to sponsor a balloon release.

6. Sponsor a second, third, fourth, balloon release. Compare the results. Keep making the point: RADIOACTIVITY TRAVELS. No matter where we live, we are not safe from this invisible pollution.

HOW TO DETECT LOW LEVEL RADIATION

One of the major problems involved in convincing large numbers of people that radioactivity is dangerous is that, unlike other industrial pollutants, it is both odorless and invisible. To the assertion that radioactivity has been and continues to be routinely released from nuclear power plants, proponents of nuclear power have always replied, "Show us the proof." In this section Japanese researcher Sadao Ichikawa explains how to use the Spiderwort plant to prove that radiation is routinely released into the biosphere from nuclear power plants. Although it requires a good deal of perseverance and hard work, the Spiderwort Strategy is not overly technical and could very easily be used by a high school science class or a grass roots organization.

Once you know that radiation is being released from a nuclear power plant the next step is educating residents of the area about the dangers of low level radiation. In the second article in this section Trumansburg Rural Alliance member Anne Burgevin suggests a few ideas for making members of the community, particularly doctors, more aware of the health hazards of low level radiation.

SPIDERWORT STRATEGY

By Sadao Ichikawa

Here is an explanation by Sadao Ichikawa of the Laboratory of Genetics, Faculty of Agriculture, Kyoto University in Japan of how the spiderwort strategy works to detect increases in environmental radiation due to leaks or routine releases from nuclear power plants.

When atomic testing began in the Nevada desert it was widely believed that low doses of radiation were relatively harmless. Government agencies had established a threshold for the amount of radiation which people could absorb on a yearly basis without endangering their health. Today, after many years of studying the effects of low level radiation on human health, scientists like John Gofman, Irwin Bross, Thomas Mancuso and Rosalie Bertell have concluded that there is no safe threshold for radiation. If absorbed in low doses over extended periods of time, radiation can cause cell mutations, cancer, premature aging and death.

That radiation causes cancer and genetic mutations is not denied by spokesmen and women for the nuclear industry. But the nuclear industry does deny that the radiation which is routinely released from nuclear power plants and the various phases of the nuclear fuel cycle is extremely dangerous. Unlike other industrial pollutants, radiation is colorless and odorless and, therefore, not easy to detect by those living near a uranium mine, enrichment or nuclear plant, or storage facility. Activists who wished to demonstrate that radiation is routinely released into the biosphere from nuclear plants had to discover some method for proving their claims.

A method for detecting radiation releases from nuclear plant sites has been used in Japan by Dr. Sadao Ichikawa and his colleagues at the Laboratory of Genetics, Kyoto University, Japan. The method involves planting hundreds of tiny flowers called spiderwort (the scientific genus name is Tradescantia) near the plant sites and periodically examining the flowers to detect color changes which are caused by low level radiation. According to Dr. Ichikawa, "This system is regarded as the most excellent test system ever known for low-level radiations."*

The color changes occur in the tiny hairs which cover the spiderwort's stamen** and are evidence that radiation

---

* From *The Spiderwort Strategy* by Dr. Sadao Ichikawa

** Stamen are the pollen-producing organs of a flower—thin, feeler-like projections which are surrounded by the petals.

has caused a cell mutation in the flower. Because the spiderwort possesses one dominant blue gene and one recessive pink gene, the change in color is from blue to pink. Each flower has six stamen and each stamen bears about 50 to 90 hairs which can be examined for cell mutations. 300 to 500 stamen hairs or as many as 8,000 to 15,000 stamen hair cells can be observed in a single flower, making it relatively easy to observe the pink mutation. The mutations can be observed about 8 to 18 days after exposure to radiation, most efficiently after 12 or 13 days. The stamen hairs are observed by placing them in drops of liquid paraffin on a glass slide and observing them under a stereoscope at a magnification of 15 to 20 times. Dr. Ichikawa explains: "The stage glass of the stereoscope should be of a milky-white color, and the light source should be of white color in order to observe the pink and blue colors. The numbers of stamen hairs and of pink mutations should be recorded for each stamen. Pink mutation is defined as a single pink cell or two or more contiguous pink cells between blue cells, and is considered to have been derived from a single mutation."*

Describing the efforts of a Japanese biology teacher to detect radiation releases from a plant near his home, Dr. Ichikawa writes:

"One snowy day in February, 1974, I had a visitor at my laboratory office. The visitor, Mr. Motoyuki Nagata, a biology teacher at Sagara High School, Sagara, Shizuoka Prefecture, asked me if the spiderwort could be used for observing radiation leaks. He was living in Sagara near the Hamaoka Nuclear Power Plant owned by Chubu Electric Power Co. which was nearly completed at the time. 'Well, honestly speaking,' I told him, I am rather pessimistic about detecting an increase of mutations under normal operation of the reactor. You may have to observe at least several million stamen hairs, and this might be too much for one person.' Mr. Nagata still wanted to try the spiderwort strategy and he asked, 'Will I be able to detect some increase in mutation frequency if the nuclear power plant discharges abnormally high radioactivity?' 'Yes,' I answered, 'You may

---

* *The Spiderwort Strategy*

but you have to be well prepared to carry out very tedious and hard work from which you may not get an answer.'

"Mr. Nagata came again to Kyoto in late March of the same year and practiced observing spiderwort stamen hairs for about one week until he became confident in his technique. On May 27th, I drove down to Sagara with 40 young plants grown from the cuttings of the same single plant. They were planted in 24 cm pots and placed at nine different points in Hamaoka (0.58 to 2.7 km apart from the reactor) and one point in Sagara (8.3 km apart) on June 2nd. Mr. Nagata bought a new stereoscope and continued practicing his observations for about one month before he began regular counts of mutated cells on July 7th. He collected four flowers from each of the ten spots, put them into small containers with moistened bits of sponge, gave the flowers water and nutrients and searched for pink mutations in the stamen hairs of 40 flowers each day. When the particular clone (Ku 7) of the spiderwort plant Mr. Nagata was using terminated its flowering season on October 31, he had examined about 640,000 stamen hairs or about 17,600,000 stamen-hair cells and had 2,778 pink mutations from this sample.

"Mr. Nagata forwarded a thick pile of data to me in which the mutation frequencies were calculated separately for each of the ten points and for eight periods of two weeks each. Each of the mutation frequencies was then cross-checked with the mutation frequency at each point *before* the nuclear reactor began operating and I soon found some statistically significant increase in the pink mutation frequency. The mutations had occurred at six out of the nine points in Hamaoka, most conspicuously at three points, and five out of the six points where the pink mutation frequency was higher were those located to the leeward side of the power plant, or *downwind*. The remaining point was closest to the reactor. The pink mutations occurred most often when the reactor was in operation.* The Hamaoka Nuclear Power Plant started its test operation on the 13th of August and was shut down on the 2nd of October because a crack occurred in one of the cooling pipes. The genetic effects of radiation exposures appear on the stamen hairs most conspicuously 12 to 13 days later.

"The results of Mr. Nagata's and other teachers' work seemed to verify a correlation between the release of radiation from the nuclear plant in gaseous form and the pink mutations in the spiderwort. But before we could conclude this we had to examine other possible factors in the environment such as pesticides, fungicides, insecticides, herbicides, automotive air pollutants, natural radioactivity, fallout due to atmospheric testing, temperature variation, rainfall record, etc. We were unable to make any significant connection between these pollutants and the pink mutations in the spiderwort.

"The factor causing these mutations is gaseous radioactive substances which are discharged from nuclear power plants.

---

*This coincides with Dr. Ernest Sternglass's research in which he found significant increases in infant mortality and cancer rates *after* a nuclear power plant began operating in Connecticut.

They are predominately radioactive rare gases such as krypton-85 and xenon-133. Besides these rare gases, iodine-131 with a half-life of 8.14 days is discharged at rates between one thousandth and ten thousandth of rare gases and is very significant because it is known that iodine-131 is concentrated in plant tissues--with an extremely high concentration factor of 3.5 to 10 millions. Cobalt-60 and manganese-54 are also released into the environment through chimneys and these nuclides have been found in pine needles.

"The environmental radiation levels in the area surrounding the Hamaoka Nuclear Power Plant have been continuously measured by the Health Institute of Shizuoka Prefecture and the Chubu Electric Power Co. I analyzed the data and found the radiation level had been increased as much as 7.5 and 8.7 millirem per year in 1974 and 1975 after starting the test operation of the reactor. Because of the nature of the dosimeters used to measure the radiation in this area only increases in gamma rays were documented, but the increases were far in excess of the less than 5 millirem increase per year guaranteed by the plant's operators. In spite of our scientific observations the prefecture and the power company continued to tell the public that they did not detect an increase in radiation.

"The increase in radiation level was only 7.5 or 8.7 millirem per year, but this increase can have a significant effect on the health of area residents because: some radioactive nuclides are concentrated greatly in the biological systems--for example, iodine-131 mentioned above; when radioactive nuclei attach to plants or animals or are incorporated into their tissues, absorbed radiation, especially of beta rays, becomes very much higher. In other words, the increase of environmental radiation level monitored actually represents only a part, and most likely only a minor part, of the actual dose of radiation which may have been absorbed by living organisms.

"The method of monitoring workers' exposure to radiation is also extremely inefficient. The exposure data comes from the readings of pocket chambers of film badges which are measures of external exposures only. These workers often spend time in contaminated areas where many kinds of radioactive nuclides are floating in the air and attaching to the walls, floors, tools, machines, handrails, etc. Internal exposures cannot be avoided and these internal exposures may be more significant than one-time external exposures. Spiderwort taught us much about the effects of radiation that had been previously ignored."

Since Mr. Nagata's observations of the spiderwort, this method of observing low level radiation releases around nuclear power plants has gained world wide attention. The lovely flowers of the spiderwort, which are honest enough to show radiation effects within only one or two weeks, are continuously sending *stop signals* to nuclear power by changing their color from *safe* blue to *dangerous* pink. The spiderwort strategy, though it requires hard work and persistence, can be used by grass roots groups throughout the world to demonstrate that nuclear power plants do increase the level of radiation.

BEYOND THE SPIDERWORT STRATEGY

By Anne Burgevin

After reading the Spiderwort Strategy, Trumansburg Rural Alliance member Anne Burgevin began her own research into the dangers of low level radiation. Here are a few of Anne's suggestions on how to convince your doctor, dentist or local legislator that low level radiation is extremely dangerous.

In his book *Irrevy: An Irreverent, Illustrated View of Nuclear Power,* John Gofman asks: "Would we be willing to accept nuclear power if we had to name 100 or 1,000 or 100,000 people each year to be executed by firing squad in exchange for electricity? Is this any different from what we are accepting, if we permit nuclear power to go ahead?"* That radiation causes cancer and genetic mutations is well documented, yet the nuclear industry continues dumping radioactive waste into our rivers and streams and releasing radioactive isotopes into our air. For example, at the federal waste dump in Hanford, Washington, hundreds of thousands of gallons of high level liquid waste have leaked into the ground and may one day contaminate the entire food supply of the Pacific Ocean. In New York state, radioactive isotopes from the former reprocessing plant at West Valley have been found as far away as the Ontario Basin.

One of the most cynical arguments in favor of nuclear power is that "no one has ever died from nuclear power." But this argument does not stand up under scrutiny by radiation researchers investigating cancer rates among the Navaho uranium miners, workers at the Hanford, Washington waste depository and shipworkers at the Portsmouth Naval Shipyard.

In the southwestern United States Navaho Indians have been working for years in poorly ventilated mines owned by Kerr-McGee, the company for whom Karen Silkwood worked until she was killed in a car accident.** The miners inhaled radon gas while they worked and, according to the Union of Concerned Scientists, deaths from lung cancer among the total group of some 6,000 miners is currently 250-300, while predictions of 600 to 1,100 excess lung cancer deaths have been made for the miners. Studies of workers at the Hanford, Washington waste storage site and the Portsmouth shipyard have also shown a high rate of death from cancer:

"Recent epidemilogical studies have begun to reveal the tragic consequences of occupational exposure to radiation

hazards. In an examination of over 100,000 death certificates, including 1,752 from Portsmouth Naval Shipyard workers, Dr. Thomas Najarian recently made a horrifying discovery. While national mortality from cancer was 18% and that of shipyard workers was 21%, fully 38% of shipyard workers who worked on nuclear-related projects (with a risk of radiation exposure) died of cancer.*

For every year that nuclear power is allowed to continue, thousands of men, women and children will die slowly and painfully from cancer, the genetic pool will mutate and the land poisoned for hundreds, even thousands of years.

As a woman who would like to give birth one day to healthy children, I have asked myself many times what I can do to alert people to the dangers of radiation. Here are some of the ideas I have come up with:

Talking to doctors and dentists

Many doctors know little about how the nuclear fuel cycle works, and some prescribe x-rays all too frequently without considering the serious harm this diagnostic technique may do to the patient. It is important that we make physicians and dentists aware of the dangers of low level radiation by refusing to be x-rayed until it is proven *beyond a reasonable doubt* that this is necessary. Because doctors are not accustomed to being questioned about medical procedures, your argument should be buttressed by evidence concerning the dangers of low level radiation. Three studies you might use are:

1. Dr. Irwin Bross's Tri-State study of diagnostic x-rays. In a study of 13 million people Dr. Bross found significant genetic damage and large increases in leukemia for both children and adults who had received diagnostic x-rays.

2. Dr. Alice Stewart's Oxford research which, according to Dr. John Gofman in testimony during a suit brought against the Nuclear Regulatory Commission, "has proved beyond statistical doubt that for fetuses irradiated by just a very small dose of diagnostic x-rays in the third trimester of pregnancy there is about a fifty percent increase in the inci-

---

*John Gofman, *Irrevy: An Irreverent, Illustrated View of Nuclear Power* (San Francisco: Committee for Nuclear Responsibility, 1979), p. 123.

**Many people believe that Karen Silkwood was murdered because of her investigation into working conditions at Kerr-McGee.

*Anna Gyorgy & Friends, *NO NUKES, Everyone's Guide To Nuclear Power* (Boston: South End Press, 1979), p. 194.

dence of cancer, cancer fatalities of all types, leukemia of all types, during the first ten years of life, just from the amount of radiation received from a diagnostic study of the mother; and for fetuses in the first trimester, the sensitivity is something on the order of ten to fifteen times as high."*

3. Dr. Ernest Sternglass's research which demonstrated that rates of cancer and infant mortality have increased in communities near to and downwind from nuclear reactors.

Sometimes doctors are more willing to listen to another professional than to a "layperson." But there are a number of ways we can lobby our physician or dentist to convince him or her that radioactivty, even in small doses, is dangerous to human health. Here are a few suggestions:

On your next visit to your family doctor or dentist, ask what he or she knows about nuclear power and the effects of low level radiation on human health.

Offer your dentist or doctor a copy of *Nuclear Madness* by Helen Caldicott. Dr. Caldicott is a pediatrician at Children's Hospital in Boston and she calls nuclear power "the ultimate medical insanity." Dr. Caldicott warns that if we do not stop building and operating nuclear power plants we face an epidemic of cancer in America.

Invite your dentist or doctor to a meeting of your grass roots group and show *I Have Three Children of My Own*, a fifteen minute slideshow in which Dr. Caldicott discusses the dangers of plutonium and radiation and expresses her outrage that increasing numbers of children will die "painfully and bleeding from every orifice" due to radiation from various phases of the nuclear fuel cycle.

Include copies of studies by radiation researchers with your medical and dental bills.

Keep a file of newspaper articles documenting the periodic leaks and "approved releases" of radioactive materials from nuclear sites. While your son or daughter is getting a tooth filled you can talk with his or her dentist about the latest release of radioactivity from a nuclear plant and what this could mean in terms of human health.

If your dentist or doctor refuses to believe that low level radiation is dangerous and continues to prescribe x-rays frequently I would recommend changing dentists or finding a new physician.

Vigils

Many people on both sides of the nuclear power issue believe that *facts and only facts* are needed to persuade someone that nuclear power is dangerous or beneficial. If we are to be effective opponents of the nuclear power industry there is no doubt that we must know a great deal about the economics, politics and health hazards of nuclear power, but I believe that ultimately nuclear power is a moral issue which can not be resolved by academic debate or intellectual prowess. To the widow of a Navaho uranium miner or the mother of a child dying of leukemia, nuclear power is a highly emo-

tional issue. As a woman I feel outraged that my children may be born deformed or die an early death due to the radiation released from nuclear power plants. And I would like to join with other women (and of course men who wish to join us) in performing vigils:

I would like to see groups of women perform "die ins" or "mourn ins" in front of utilities that own nuclear power plants. We would be mourning for those who have died and who will die from radiation poisoning, and by performing "die ins" we would be demonstrating in a very direct way our opposition to the *random premeditated murder* being committed by the nuclear power industry.

I think we should also attend village and town board meetings, meetings of county boards of representatives, and legislative sessions at the state level to express our anger and perform "mourn ins."

To utility executives and stockholders in utilities that own nuclear power plants we should send descriptions of how children die from leukemia and we should make it clear that even though they may be profiting from this industry *their* children and grandchildren may die or suffer genetic mutations from radiation poisoning.

I am fully aware that women are not the only victims of nuclear power, but I feel that because we bear children we have a special obligation to protect our bodies from the dangers of radiation. And I believe that if men can not stop other men from building and profiting from nuclear power, then women must.

Here is a list of radiation studies:

Bertell, R. *The Nuclear Worker and Ionizing Radiation,* Address, American Industrial Hygiene Association Meeting, May 9, 1977.

Bertell, R., *X-Ray Exposure and Premature Aging, J. Surg. Oncology,* 9:4, 1977.

Bross, I.D.J., and Natarajan, N. *Leukemia from Low-Level Radiation—Identification of Susceptible Children, New England Journal of Medicine,* 287:107, 1972.

Caldicott, Helen, *Nuclear Madness, What You Can Do,* Autumn Press, Boston, 1978.

EPA (United States Environmental Protection Agency) Mango, P.R., Reavey, T.C., Apidianakis, J.C. *Iodine-129 in the Environment Around A Nuclear Fuel Reprocessing Plant* (West Valley, NY), EPA Office of Radiation Programs, Field Operation Division, Washington, D.C., October, 1972.

Gofman, J.W. *The Cancer and Leukemia Consequences of Medical X-Rays, Osteopathic Annals,* November, 1975.

Gofman, J.W. *Cancer Hazard from Low-Dose Radiation, CNR Report,* 1977-9, NRC Docket No. RM 50-3, USNRC, 1977.

Honicker, Jeannine and Farm Legal, *Honicker vs. Hendrie: A Lawsuit to End Atomic Power.* Book Publishing Co, Summertown, TN, 1978.

Ichikawa, S. and Nagata, M. *Nuclear Power Plant Suspected to Increase Mutations,* Laboratory of Genetics, Kyoto Univ., Kyoto, Japan, 1977.

*SHUT DOWN, *Nuclear Power On Trial* (Summertown, Tennesee: The Book Publishing Company, 1979), p. 27.

Mancuso, T.F., Stewart, A. and Kneale, G. *Radiation Exposures of Hanford Workers Dying from Various Causes, Health Physics,* 33:369, November, 1977.

Natarajan, N. and Bross, I.D.J. *Preconception Radiation and Leukemia, J. of Medicine,* 1973.

Sternglass, E.J. *Radioactive Waste Discharges from the Shippingport Nuclear Power Station and Changes in Cancer Mortality,* May 8, 1973.

Sternglass, E.J. *Strontium-90 Levels in the Milk and Diet Near Connecticut Nuclear Power Plants,* October 27, 1977.

Sternglass, E.J. *Cancer Mortality Changes Around Nuclear Facilities in Connecticut,* Testimony Presented at a Congressional Seminar on Low-Level Radiation, Washington, D.C., February 10, 1978.

Stewart, A. *A Survey of Childhood Malignancies, British Medical Journal,* 1:1495, 1958.

Wagoner, J.K., et. al. *Radiation as the Cause of Lung Cancer Among Uranium Miners, New England Journal of Medicine,* 273:181, 1965.

HOW TO EFFECTIVELY ATTEND SYMPOSIUMS

On June 9, 1979, I attended a pro-nuke symposium in Painted Post, New York. I wrote to the symposium's sponsor, Congressman Stan Lundine, requesting that he sponsor a less biased discussion of the problems of radioactive waste. With my letter I included information disputing the assertion, made over and over at the symposium, that the waste problem has been solved. Mr. Lundine never replied to my letter.

Often sponsored by proponents of nuclear power, symposiums are a propaganda forum for utilities and their pro-nuclear government spokespeople. The public attends these symposiums out of an interest in the issues and in good faith, only to be misled by erroneous statements and bogus statistics about the dangers of nuclear power and radioactive waste. It is imperative that we not allow scientists employed or funded by the nuclear industry to mislead or confuse the public. If we do not attend these pro-nuke symposiums many people may be deceived by the apparent truth or logic of statements which are untrue, irresponsible and even dangerous.

A DAY AT A PRO-NUKE CIRCUS

By Fred Wilcox

"A Day At A Pro-Nuke Circus" describes the symposium at Painted Post and offers a few suggestions on how to attend symposiums.

Anyone who has been involved for long in the anti-nuclear movement has encountered the argument that nuclear power is simply too complicated for the average person to understand and, therefore, decisions about our nation's energy policies should be made by politicians and their scientific advisers who "fully understand the issues."

I can think of no better way to challenge this assumption than by attending a symposium on nuclear power or waste. Listening to the experts, you begin to sense a certain arrogance: they are on a raised stage; they teach physics or chemistry at prestigious universities; and, with or without their consent, they have been made into demi-gods by our technological society. And yet, they are making statements with which you disagree, statements which, from your own research, you know will not stand up under careful scrutiny. As the symposium continues, you will feel more confident of your ability to challenge advocates of nuclear power.

On June 9, 1979, I traveled with five members of the Trumansburg Rural Alliance to a symposium on nuclear waste held in Painted Post, New York. Sponsored by Congressman Stan Lundine and advertised as a "General Overview of the Nuclear Waste Management Problem," the symposium was to include a keynote speech by Representative Mike McCormack, Chairman, Energy Research and Production Subcommittee of the House Science and Technology Committee. He had also been named Solar Man of The Year by his colleagues in Congress.

The morning's first speaker was Dr. L.D. Pye from Alfred University who began his talk by announcing: "Today I'm going to introduce you to some chemicals which may not be too familiar to you." Dr. Pye was followed by Dr. Larry Hench from the University of Florida. Dr. Hench started with a confession. His funding, he confided, came almost entirely from government sources, some of which were "directly or indirectly" military. But if this bit of honesty disturbed anyone, Dr. Hench hastened to add that his work had led him to the conclusion that all the technological problems involved in storing radioactive waste had been solved. According to

Dr. Hench, only political problems remained. Following this rather startling news,* we were entertained with a series of graphs designed to reassure us that if stored properly, meaning in glass containers, nuclear waste would be no more dangerous in three to four hundred years than uranium in its natural state. Dr. Hench used a somewhat peculiar method to prove all this: He quoted the batting averages of famous baseball players and compared these averages to the "longevity capabilities" of his newly discovered wonderglass. Though it was an interesting diversion, it wasn't clear to some of us just how Willy Mays' or Babe Ruth's batting .356 in one season proved that radioactive wastes could be stored safely in glass. When the morning session ended one of our group rushed to the stage and asked Dr. Hench if he knew the half life of plutonium. He did not.

As the morning wore on, it became more and more apparent that the charts, graphs, drawings, statistics and baseball averages were designed for one purpose: to demonstrate that through a process called *vitrification* radioactive wastes could be safely turned into solids and that, because of the unquestionable superiority of glass, which will succeed where steel, concrete, and other substances have failed, the problems of storing waste have been solved.

During the six hours the symposium lasted, no one asked who would receive the contract for all these super-glass storage vessels. No one needed to. You could almost see Corning Glass Works through the parted curtains of the motel.

Mr. McCormack began his talk by holding up a geiger counter which went click, click, click when he waved it about the room. "Now, as I point this geiger counter at the room, this is about as much radiation as you would find at the entrance to a nuclear plant." (A reassuring soft baby click click.) "But as I point it toward the bricks here," (a brick wall behind the podium) "It goes up a bit in rate..." (A slightly stronger, mother hen click.) "Because, you see,

---

*According to the *McNeil Report,* over five thousand studies of the waste storage problem have been made and a reliable method for safe storage *has not been determined.*

there is more radiation in these bricks than at the entrance to a nuclear plant." Deduction: One small click + One large click = nothing to worry about.

Among other things, Representative McCormack declared that the 500,000 gallons of radioactive waste which have leaked into the soil and water around Hanford, Washington, pose "No threat to the biosphere,"* and that New York State must develop its nuclear generating capacity or face a shortage of electricity by 1990. When the Congressman made the statement about Hanford, pandemonium broke out. People who had been listening with agitated patience shouted challenges. There was a collective groan from the back of the room and a few boos. Someone cried: "Take McCormack and his bullshit back to Washington..." A young man sitting just a few feet from Congressman Lundine moaned: "This is a fraud, a ripoff. I'm sorry I paid five dollars for this. I'd really like my money back." Congressman Lundine ordered the young man to leave but he refused. "That's democracy for you, Congressman," someone shouted. A member of the Trumansburg Rural Alliance leaped to his feet, waving a copy of a New York State Gas & Electric Report and demanding a chance to question Rep. McCormack. Rep. Lundine conferred with an aide and announced: "THIS SYMPOSIUM WILL BE CONDUCTED IN AN ORDERLY FASHION OR NOT AT ALL." The mumbling and shouting continued. "AND I CAN ASSURE YOU THAT THE LOUDEST WILL NOT BE CALLED ON FIRST."

Then Congressman Lundine called on David Curtin, a member of our group. "Congressman McCormack, are you aware that New York State now has a ten thousand Megawatt surplus electrical generating capacity, and that the demand for electricity in this state has not gone up since 1975?"

Mr. McCormack smiled, stroked his geiger counter, and said: "I'm afraid I'm not really familiar with the report you are referring to."

"It says that what you just said about our running short of electricity if we don't build nuclear plants isn't true; that we are not dangerously close to using up our generating capacity. On the contrary, Congressman, we have more than we are using right now, so why build more plants?"

There were a few snickers, some amazed gasps, a few heads bowed in commiseration for the keynote speaker who had been so amusing and entertaining. More questions were asked Rep. McCormack and his response was consistently vague or unresponsive. The lunchtime diversion ended with more praise for "Mike," and many thanks for his kindness in taking his valuable time to come all the way to Painted Post to address this "vital matter." After a fifteen minute break during which we struggled to regain enough composure to face the afternoon session, we took our seats and prepared to listen politely to "experts" discussing the world's future.

Among other bits of technological wizardry, we heard

an employee of Corning Glass say that *vitrification* would turn radioactive wastes into ceramics and would solve much if not all of the problem of what to do with 200 million gallons of high level waste, 400 million cubic feet of low level waste, and 95 million cubic feet of alpha waste which will have accumulated by the year 2000 A.D. from federal and commercial reactors.

Our patience had nearly run out by the time Dr. Robert Pohl from Cornell University was allowed to speak. Dr. Pohl began by challenging the idea that radioactive waste buried for three or four hundred years in glass containers would be no less dangerous than uranium in its natural state. Dr. Pohl explained that with the passage of time these wastes become intensely hot, gases form, and pressure builds up which cracks containment vessels. One crack is enough to release dangerous isotopes into the biosphere. The technological problems of solving radioactive waste, Dr. Pohl warned, have *not been solved.* Unfortunately, we had to wait several hours for Dr. Pohl and by the time he spoke the audience had not only been exposed to a great deal of questionable information and a noon hour sideshow, but many people had left—including Representatives Lundine and McCormack and most of the morning's speakers.

Unable to express our disappointment, we left Painted Post depressed. Having been active in the anti-war movement for many years, symposiums or public debates were not a new experience for me. Nevertheless, it left me drained, feeling anxious, surprised at my own naivete. Once again we were listening to "experts" discuss *our* fate and the *fate of the world* with a dispassion which was supposed to inspire confidence because, we were told, dispassion is the mark of a true scientist. Once again we were listening to "experts" make foolish or erroneous statements. And if someone challenged these statements, the moderator would threaten to call the whole thing off or evict the "disruptive element."

In Painted Post we were unwilling to risk a confrontation and, therefore, waited hours for a chance to refute the numerous false statements made during the symposium. But we waited in vain because the symposium was structured so that the propaganda would have the maximum effect. In the future we intend to be better prepared for and far less patient with the pro-nuke apologists.

Here are a few suggestions that may be useful in preparing for a symposium on nuclear waste, the transportation of radioactive materials, or the future of nuclear power.

1. Try to find out who the speakers will be and do research into their backgrounds. Find out where they teach, if they have done research into nuclear power, where they get their funding, if they work or have worked for the utilities, what connection they have with the sponsor of the symposium, what papers or books they may have written on the symposium's topic.

2. In order to avoid a shouting match with the speaker who is making false statements at a symposium, you could make

---

* "The Federal Water Pollution Control Agency charged that the AEC's negligence at Hanford 'had resulted in the world wide acknowledgement of the Columbia River as the most radioactive river in the world.'" —John J. Berger, *Nuclear Power: The Unviable Option.*

posters before hand with quotes about the dangers of nuclear waste--or whatever the day's topic may be. Each member of your group could take a poster to the symposium and, if a speaker is insisting that the waste problem has been solved, nuclear power is safe or radioactive materials can be safely transported, you could hold the card so everyone in the room would be able to read a statement contradicting the speaker. If the press is attending the symposium your chances of getting coverage this way are good and your silent protest against pro-nuke propaganda may reach a wide audience.
test against pro-nuke propaganda may reach a wide audience.

3. If you know the symposium is going to be a pro-nuke circus, find out if you can show a slide show or film in a nearby lobby or room. And even if you do not get permission to do this, arrive early and distribute fliers about the dangers of nuclear power and the storage and transportation of radioactive wastes.

4. Record as much of the symposium as you can so that you will have direct quotes when you write letters to the editor of your local newspaper challenging the veracity of what was said at the symposium. You can also use these quotes when you write to the symposium's sponsor demanding a follow-up to the symposium, and when you send the sponsor information which refutes statements made at the symposium.

5. Write to the speakers who made pro-nuke or erroneous statements and tell them that you disagree. Send them information which disputes their statements and a bibliography which they can use to do more research into the topic. Suggest they come to a future, less biased symposium.

HOW TO OPPOSE NUCLEAR POWER WITH MUSIC

From the days when the International Workers of the World marched into towns all over America singing "Halleujah I'm A Bum" to the Civil Rights Movement when Martin Luther King led thousands of followers into Selma and Birghamton singing "We Shall Overcome," music has been an important part of any social movement. Anyone who has attended even one demonstration knows that crowds become restless when too many speakers make the same point. But when energies are waning and patience wearing thin, a song or two can inspire the crowd once again, renew energies and send people away even more determined to work for a better world.

TRAVELS WITH AN ANTI-NUCLEAR MUSICIAN

By Court Dorsey

In this section Court Dorsey, musician with "Bright Morning Star" and anti-nuclear activist who was arrested in 1977 for opposing construction of a nuclear plant at Seabrook, New Hampshire, explains why music is a powerful tool for social change and offers a few suggestions for using music to oppose nuclear power—and for building a new and brighter future.

Camping At The NRC

Following a week of demonstrating against the construction of a nuclear plant at Seabrook, New Hampshire in 1978, we gathered on the steps of the nation's capitol to continue our protest. Calling ourselves the Seabrook Natural Guard and consisting of a hastily formed coalition of affinity groups, we were in Washington to pressure the NRC to suspend construction on the Seabrook plant. We listened to speeches by Congressmen Toby Moffit and Leo Ryan and songs by two of New England's musical activists, and then we began our march to the Nuclear Regulatory Commission, singing as we went. At the end of the March there was a larger rally with more singing and speeches. The speeches and songs complemented one another, the speakers outlining issues and demands and the songsters calling to people's hearts.

We spent 4 or 5 days on the sidewalks in front of the NRC, baking in the sun, soaking in the rain and sleeping on the concrete after an exhausting day singing to passersby. To people in the street we sang about why we were in Washington, to the police we sang about disarming and humanizing our culture. As the days passed and tensions increased, our singing held us together when we were near the point of total exhaustion and frustration. And when several members of the Nat ural Guard were arrested and carried away for "dying in" in front of the NRC's doors, we sang a song of rejoicing to our friends; we sang encouragement and our sorrow over the continued use of nuclear power.

I'll never forget the image of the eighteen who occupied the NRC building, refusing to leave until plant construction was ordered suspended, framed behind a window of sound-proof glass which separated them from those of us still on the street—how they stomped and swayed and rocked inaudibly behind the glass, while we sang and danced with them, eye to eye, songs that we all knew so well that we *could almost hear each other.* We made many friends during those days, not on the Commission itself perhaps, but in the streets and among the employees of the NRC. And when the decision was announced to suspend construction (temporarily, of

course) there were statements by the occupiers, and a victory dance in the street. Music was a part of each phase of the action.

Music is a unifying power, the power of the many finding themselves in the one, the whole. To discover this one need only listen. If one is with others and each makes a sound, their voices will blend into one sound, one voice. Try this in a circle, facing the center, on a single common sound; as you listen you will hear one voice, and you will feel the unity in and among you all. Sight tends more to show us distinctions, where one leaves off and the next begins. Not so with sound; sound tends to show us where we blend, where we are one. Is it any wonder then, the sense of empowerment that comes from singing or chanting with others?

Music is inherent in movements for social change. My friend and fellow musician Charlie King has thought a good deal about music within this context and has concluded that, "Any viable people's movement has its own music. Whether a movement has a good lively musical tradition is an indicator of the health of that movement." For example, Charlie compares the active, militant unions that sing the old union songs at meetings to the older and more bureaucratized unions that have to hire somebody to sing to them at a banquet. "If there is a people's movement," Charlie says, "there will be singing because the newspapers don't always tell the whole story. If people want to share their story, if they want it recorded, they're going to have to put it into song." In this sense music is a living history, adding our own story to the tradition of working for social change. Music helps us gain perspective, enables us to see ourselves in a greater context, and teaches us patience. Through music we gain wisdom, and through social action our songs are recorded for posterity.

Music is an important tool for social change which organizers would benefit from using well. Nothing can be more counterproductive than an afternoon of redundant political speeches. This is an invitation to failure because, regardless of how vital the messages may be, there are limits to the amount of information people can absorb. Between speeches we should enjoy music, singing, even dancing. The songs will regain exhausted attention and energize the group, inspiring people to think about why they are at the rally. The music will transform the barrage of facts and statistics into a celebration of life and strengthen solidarity by expressing the joy we feel working together toward a common goal—in this case the elimination of nuclear power. Music can resolve impasses and break through deadlocks at meetings, alleviating the tensions and exhaustions of prolonged strategy sessions. Indeed, one of music's most important properties is its healing power.

Singing to the National Guard at Seabrook, New Hampshire

In the spring of 1977, I traveled to New Hampshire with an affinity group of activists from DeKalb, Illinois, to take part in the non-violent occupation of the Seabrook Station construction site. Most of the group with whom I was traveling were musicians and actors so we arrived early to do street theatre and music for a few days before the action. Rolling into town, we would find a prominent spot to perform, passing out leaflets explaining our opposition to nuclear power and our support for the New Hampshire seacoast residents in their efforts to halt construction on Seabrook Station.

Later we were arrested with 1400 demonstrators and incarcerated in National Guard armories around the state. During our two week confinement we had considerable opportunity to sing and perform our theatrical pieces. Music, poetry, and other arts were an important part of the cultural atmosphere of armory life.

Affinity groups were the cornerstone of our organization, and though they could be re-formed if broken we wanted to keep them intact. So when our jailers told us they intended to separate the men from the women and to send the women to another armory, we decided to resist. After many hours of meetings (the guards could see but not hear us) we each decided how our affinity group would respond, removed our boots and shoes, tied all 150 pair together in a huge pile in the middle of the floor and sat down to watch a play. When the play ended we were informed by our guards of a change in plans: Any seventy people could leave to decrease the population in that particular armory. We could choose. We needn't break down our groups. We agreed.

The space in which we were being held was a gymnasium, and the day after our arrest we were told that the women and men would be separated by partitions. Again we held meetings, but while no unanimous agreement could be reached, we decided to support one another when the move to separate us came. I remember vividly the CO as he and his reluctant but determined line of men charged in, shouting in the most authoritarian manner, "All right! It's comin' down!" They proceeded to tear apart locked arms, partition off one corner of the space with two lines of rope, drag men from the women's side and women from the men's side (just to have them duck back under the ropes again), getting rough at times and threatening, but many of them obviously hating

what they were doing. Finally there were two rooms, separated by two ropes with a line of Guardsmen inbetween. Some demonstrators were weeping, and the guardsmen were struggling to avoid eye contact. Spontaneously the men on one side of them, the women on the other, joined hands and began softly singing "The Love Chant," a simple melodic round that everyone knew:

> Love, Love, Love, Love
> People are made for love
> Love each other as ourselves
> We are one.

I remember reciting one of the speeches from our street play about the dangers of nuclear power. Still chanting, others in the gymnasium began facing and speaking directly to the guardsmen. The spirit of the music permeated the contact, and the men in the middle were deeply moved, some even weeping, and each man had to be relieved every twenty minutes in order to regain his composure. They were not allowed to speak. They were following orders that for the most part they didn't like. After three hours of continuous singing and talking, we agreed to the CO's proposal that the ropes would be up only from midnight to 8:00 a.m. During the day the room would remain open.

Yet music is more than a unifier. Take a song. It is an expression of the whole person. The heart or passion of the singer is the wing of the song. Her soul or spirit is the air in which it has life. The lyrics spell out and communicate her concerns and give information about her relationships with the persons, thoughts, or things of her life. The singer's body, her physical being, cooperates in bringing the song about. The will is actively involved in tuning and balancing harmony, rhythm and pitch—at once abstract and pulsating. In short, the song is an integrated expression of a whole person; it is a unifier within the self. Is it any wonder then, the sheer delight that accompanies the sensitive hearing or making of music? One experiences the joy inherent in the intuition of the universal, which is true revolution.

Songs are a powerful tool for building networks and coalitions. A performer's or performing group's repertoire can aid in coalition building as well. Singers and songwriters often travel around the country learning songs from different struggles. Many see the beginnings of a social change towards peace, a clean environment and a healthier, happier human family. When people hear songs about the changes taking place in our culture, they often identify their own story with the lyrics and they feel less alone. They discover a shared concern and that sharing is fertile ground for future coalitions.

To give an example of how this all works, one week late in the fall of 1979, I and other members of the group "Bright Morning Star," with whom I sing, were on tour. We sang and gave workshops on Tuesday and Wednesday for a "Human Rights Week" at a community college in Rochester, New York. The workshops covered many topics, including the mid-east peace movement, Native American sovereignty struggles, nuclear disarmament and the anti-nuclear movement, all presented in a human rights perspective. By Thursday night we were on the Canadian border with the Mohawk Nation at Akwesasne. We had heard a good deal about their difficulties, but we wanted to listen and understand even more. We shared our music with them, singing about Native American struggles, sexism, labor disputes and anti-nuclear events. And they shared their music with us, and we danced late into the night. By the weekend we were singing at a labor rally marking the twenty-third week of a local strike in Bridgeport, Conn. We sang a song about a United Mine Workers' strike, and one about opposition to nuclear power, ending with *Keep Your Eyes On The Prize, Hold On.* That evening and the next we were at a campus coffeehouse doing concerts sponsored by an anti-nuclear alliance. Each group that week heard and understood the concerns of all the others, and we took something from each experience and carried it on.

In our present world situation the truth that the earth and its peoples are integrated parts of one whole ecological system has become painfully apparent. It is not enough that we have attained some degree of personal integration, or merely some familial or national integration within our limited group. This same principle of integration must extend through the very fiber of all existence, must be that fiber, and living in abuse of this truth we have led ourselves to the brink of extermination—the air we breathe and the water we drink are poisoned and testify to our being more interested in profit than peace. Pollution and war demonstrate that we are not at peace with nature, with the whole of which we are a part, in which we live and have our being. When this consciousness dawns in the singer, then the song takes on an added beauty. Its musical power is increased by the singing of the rocks, the trees and the sky, as well as by the silent sound of peace that dwells deepest in the human soul. And the music of such a singer has the power to arouse and nurture that spirit in others, and to bring a deep sense of celebration and joy to those already working to help bring that worldwide wholism or integration into focus. Music is like celebrating the final victory in the midst of struggle.

Using Music For Grass Roots Fund Raising

A concert of songs is an instrument for reaching out to the community beyond the movement. A good song can touch young and old alike. And a well organized concert can help raise funds for grass roots opposition to nuclear power. The first thing to remember, if you want to make money, is to keep your expenses low, rely on volunteer help as much as possible, and try to find a good local group with a substantial following. A performer or group from out of town whose music may convey the spirit and political content you want will mean higher expenses and possibly a smaller audience. With a local group you can avoid having to pay travel expenses and the anxiety of worrying about whether the group will

show. Selling tickets to a grass roots concert can be fun when you use this simple method: Simply assign each member of your group a number of tickets and have them sell them to friends, neighbors, parents and anti-nuclear sympathizers. For free concert space check with local churches. Many churches are becoming more socially conscious and are willing to allow groups opposed to nuclear power to use their facilities for meeting places or concerts. You can spend a few days baking cookies and other good things to eat, set up your literature table and a "hat" to accept donations, and make sure you have an intermission where people can talk, meet the musicians, check out the literature table, buy your baked goods and make contributions to your organization.

If you want to attempt a big fund-raiser with a well-known performer, I would suggest contacting someone who has experience with large concerts. There are innumerable pitfalls and many necessary details to look after as concerts get larger. The expense of mounting a concert like this can be substantial. You stand to make a bundle if you are successful, and you could lose your shirt. Even something as unforeseen as another event in your area on the same evening can throw off ticket sales. There are expenditures for large hall rentals and performer expenses are likely to be higher, even for a benefit. You will probably need to rent a sound system and a crew to run it, hire a rent-a-cop, and for a large concert there will be investment in advertising. It is of little or no value to spend much money on advertising beyond postering if the group is not well-known in your area. Any free spots, such as public service announcements or entertainment page stories (if you can get them) should be used. Or your group may interest some local paper in covering your concert as a news event, particularly if it is building towards a demonstration, or as part of an overall campaign. This doesn't replace advance ticket sales, because you have no idea if it will work. But it's worth a try. In a large concert, more formal and costly advertising will probably be desirable. Try smaller, more manageable concerts at first. Then, when you've developed your talent for staging larger concerts, go right ahead and plan a big event—keeping in mind that bigger is not necessarily better.

One final suggestion in this area which you might try is to work either with or through universities, colleges, or community colleges in your area. They usually have money and/or facilities available, and you might interest some faculty member or student group in co-sponsoring something with your organization. Or you might arrange for a traveling group or performer to do a higher paying concert at a local college to cover expenses, and then put on a less expensive benefit concert for your own group near the same time. Traveling groups usually prefer to do several concerts in a given area anyway, and since travel expenses are often a large factor in the cost, if you can arrange to have those expenses covered by a group with more money it may lower the cost for the benefit concert. Often smaller colleges and local community colleges can band together to share costs. Just keep in mind that the more groups involved the less money your group will be able to keep. Anything you can work out is fine.

Bright Morning Star received a letter not too long ago from a friend who is serving time in a federal prison near Rocky Flats, Colorado. He is in prison because he can not accept the continued production of nuclear bombs. He told us how important many of our songs have been to him during the months he has been confined and how he feels extremely close to us and to the others with whom he has shared those songs. He writes that he sings them often, aloud or inwardly, and they invariably remind him that his action was taken in the name of peace.

The song, in springing from the deep well of the whole, touches what is limitless, what is beyond ourselves, our prison and our death. And so a song is power. When it is sung no tyrant can destroy it. A good song sings itself even in silence in the hearts and minds of those who are alive with the life from which it springs.

Following the coup which overthrew Salvador Allende in Chile, the Chilean singer Victor Jara who had sung so untiringly for the liberation of the Chilean people, was arrested with thousands of others and held in a makeshift prison. Guards gave Victor a guitar and asked him if he would like to sing. So he began singing and they smashed both his hands. But he kept on singing and the people in the stadium sang with him until he was shot by the guards and died. Some say he still sings. We know his songs live on.

HOW TO USE TELEVISION AND THE PRESS TO OPPOSE NUCLEAR POWER

In "The Show and Tell Machine" * *Cornell Professor Rose Goldsen writes that the power of television to influence public opinion is tantamount to mass hypnosis. After the debacle of Three Mile Island, the utilities began to mount a media blitz to hypnotize the American people into once again believing in nuclear power. To counter this, we must learn to use the media by making public service announcements, writing effective letters, appearing on television and radio programs and cultivating good relations with the press at demonstrations and civil disobedience trials. We must reach the American public with our message that nuclear power is both dangerous and unnecessary.*

---

*Rose Goldsen, *The Show And Tell Machine, How Television Works And Works You Over*, (A Delta Book, 1975).

**Not Man Apart*, published by Friends of the Earth, San Francisco, California.

A 1.6 MILLION MEDIA BLITZ

This article, taken from Not Man Apart, *November, 1979, should serve as both a warning and an inspiration to anti-nuclear activists. The media blitz is on and we must prepare to meet the utilities' blitz with one of our own.*

One way the industry plans to make a comeback is by strengthening its public image, badly wounded by Three Mile Island and the subsequent credibility fallout. The nuclear industry has just launched a massive—$1.6 million—media campaign to reassure Americans that it means to prevent another Harrisburg.

Jack Horan of Knight News Service, reports that:

● Teams of utility executives will spread across the country to hold press conferences and appear on TV talk shows in Los Angeles, Boston, Seattle, Cleveland, Phoenix, Dallas, Kansas City, and, tentatively, New York, San Francisco and Atlanta;

● Starting in November, 1979, pro-nuclear advertisements will appear in magazines aimed particularly at women buyers; and

● Videotapes of experts discussing technical aspects of nuclear power will be distributed free to TV stations and information packets will be sent to the print media.

● Nuclear Energy Education Day (NEED) will be celebrated on October 18, 1979, with "more than 2000 varied events from a brunch for congressional wives in Washington to a joggers' mass relay race in California."

Jack Betts, director of the campaign, said the effort is needed to convince the country that without nuclear power the country faces an energy crisis in the 1990s. (This is what Mark Hertsgaard calls "energy blackmail.")

Two nuclear engineers will follow Jane Fonda and Tom Hayden around the country as a "truth squad" to refute their anti-nuclear arguments, Horan learned.

Betts said the money to finance the media blitz has come from nuclear plant manufacturers, investor-owned and public utilities, and industry trade organizations. A representative of Westinghouse confirmed that it has contributed to the PR campaign, but would not say how much.

Richard Pollock of Critical Mass, an anti-nuclear group in Washington, DC, said he has heard rumors that the utilities had collected around $25 million for the effort. "We are anticipating the nuclear community will attempt to polish its image with a new propaganda campaign the like of which we have not seen since the Atoms for Peace Program [of the 1950s] ."

SIX ARTICLES FROM MEDIA ACCESS PROJECT

The next six articles are from Media Access Project, a non-partisan public interest law firm and educational institution which works to insure that the media fully and fairly inform the public on important and controversial issues. MAP has appeared before the Federal Communications (FCC) and other state and federal agencies and the courts on behalf of hundreds of national and local organizations which work on civil liberties, environmental, consumer, women's and minority issues.

For more information, write Media Access Project, 1609 Connecticut Ave. N.W., Washington, D.C. 20009.

One way that grass roots groups can counter the pro-nuke blitz is by using Public Service Announcements to inform the public about the dangers of nuclear power. PSAs are short, 30 to 60 second, announcements that are aired free by broadcasters. In this section Media Access Project explains how to write and use Public Service Announcements.

PUBLIC SERVICE ANNOUNCEMENTS

One of the ways people can gain access to the broadcast media is through the format of the "public service announcement" (PSA). PSAs are short messages aired free of charge by broadcasters announcing community events or, more importantly, speaking to important community issues. They may be anywhere from five seconds to five minutes in length. Usually they are 30 or 60 seconds, because they are used in unsold advertising slots and most advertising comes in 30 or 60 second chunks. PSAs may be in the form of scripts read by an announcer--called "live copy" or "announcer copy"-- or they may be pre-recorded and elaborately produced messages on film or tape.

The Federal Communications Commission doesn't say that stations must air PSAs. But all stations have to promise the FCC that they will do some amount of public service programming in order to get their broadcast license renewed by the Commission every three years. News specials, talk shows and documentaries are some of the ways stations can fulfill their obligations for public service. PSAs are a good way for broadcasters to satisfy their obligations, since they are easily scheduled in one-minute segments (like commercials); they can represent a diversity of opinion from the community, and they are written and produced by the community for the stations; they don't cost the broadcaster a nickel to create. So almost every station promises the FCC that it will air a minimum number of PSAs each week. Once they've promised to do it, they are locked in; they must fulfill their obligation or risk losing their license.

Public service announcements are often aired in unsold air time. This means late at night, or early Sunday morning (re-- ferred to as "ghetto hours" by the broadcast industry because

fewer people are listening). But broadcasters also air a mini-- mum number in prime time as well. The total amount of air time given to a PSA depends on the station. Some are aired once a day for a week. Some are aired twice a day for four or five months. It depends on the station's public service director.

CREATING YOUR OWN PSA

We'll assume that you've analyzed your needs and decided that you want to get your message on radio and TV stations in your area. The next step is to survey the stations in your area, and find out their requirements for PSAs. For example, some radio stations require PSAs to be 55 seconds in length, so they can insert a "tag" on the end crediting themselves: "Brought to you as a public service by this station," or some- thing like that. Some stations have standing policies against pre--recorded PSAs and will use only live copy. A few sta-- tions will air PSAs which are no longer than 10 seconds in length. Once you produce your PSAs, you'll be able to ne-- gotiate with the stations as to what they'll use, but you should be aware of general parameters before you undertake produc-- tion. Simply call each station you'd like your message to ap-- pear on and ask to speak to the Public Affairs Director or to whoever schedules PSAs at the station. In many metropolitan areas there are professional associations of Public Affairs Dir-- ectors, and many of these associations publish guidebooks for PSAs for all stations in that area. Ask the first Public Affairs Director you speak to if such a guide is available. If they don't know what you're talking about, or if they think such a guide is a good idea and would like to see an example, have them write to Ms. Jane Morrison, c/o KNBR RADIO, Grosvenor Plaza, San Francisco, CA, 94102, and ask for a copy of the "Bay Area Guidebook for Public Affairs."

Once you know what the general requirements for PSAs are, you can decide what format is most appropriate to your cause. If you don't have any money to spend, you'll probably want to stick with live copy, although even with no money it's possible to do a more involved production. Basically, you want to do the best production your time, resources and out-lets allow. That's how you can reach people.

LIVE COPY PSAs

Live copy PSAs are basically the same for TV and radio. You write out your message, time it to the lengths required by the stations you want it to appear on, and station announcers read your message on the air. TV stations will use some kind of visual to accompany the reading, usually just a static slide with your organization's name and address on it. TV stations will usually prepare these slides for you at no cost or for a small token fee. Some TV stations will agree to run more than one slide if your organization supplies them. Standard color transparencies or anything that will lend visual impact to your message are all that's required. Radio stations will often ask that you supply live copy in varying lengths, from 10 seconds up to 60 seconds so that they may use it to fill whatever holes arise in their programming.

To write a live copy PSA, first figure out what the one most important thing you want to communicate is, and then state that idea as directly as possible. One statement and your name and address will give you about 10 seconds of copy, and you can then expand it from there. The most imporant consideration to keep in mind is that if you try to communicate too much, you will wind up communicating nothing. In evaluating how your copy works, don't think in terms of someone sitting in the room with you listening as you read. Think of someone driving home from work, fighting freeway traffic, getting over a hard day and looking forward to dinner, concentrating on a hundred other things when your message comes over the broken, static-filled two-inch speaker on the car radio. This is the person you'll be communicating with. Write with that in mind.

If you want people to write to you or call you, it is helpful to state the address or phone number more than once. Saying it more than twice may sound obnoxious, but in some cases it may be required—if, for example, the overriding, primary purpose of the PSA is to get people to call or write. Keep your address as simple as possible. If you can get a Post Office Box number, that'll help. In many cities it's possible to get a catch phrase or the name of your organization to qualify for a P.O. Box. For example, you can simply say, "Write ENVIRONMENT, San Francisco 94109." Talk to the local Post Office and see if this service is available in your area.

Slightly above the live copy PSA in terms of production value is the "talking head" PSA. Instead of having a station announcer read your copy, a member of your organization is allowed to use the station's facilities to tape a reading. In the case of television, your spokesperson appears on camera reading the copy.

TRY PRODUCING YOUR OWN

Most radio and TV stations will allow you to bring them pre-recorded PSAs produced by your organization. This allows you to put your message across in a more dramatic, interesting and compelling way, and to use various production techniques to enhance the communication's impact. Producing your own messages is easier than you think, but stations are very particular about the technical quality of what they put on the air. So, whatever else you do, don't cut corners on technical quality. Make sure whatever you do meets the station's technical standards—you can check with them ahead of time and find out what those standards are. If the Public Affairs Director can't give you the information you need, ask to speak to the Chief Engineer. Some stations will allow you to use their facilities to produce your own PSA, and will even provide technical assistance. If you can't get access to station facilities, check into other facilities in your area. If what you're doing can be considered a "good cause," it may be possible to get production facilities at reduced rates or even for free, if the owner of the facility is sympathetic, or just wants to do a good deed. If your organization is tax-exempt, the owner may be looking for a good tax deduction. If there are colleges or trade schools, or even high schools in your area that have radio-television classes, they may be willing to handle production for you at little or no cost, to provide their students with on-the-job experience. You may be able to find working professionals or talented amateurs—recent film school graduates, for example—who need real spots to put on their demo reel which will help them get future work. Most advertising agencies also devote at least some of their time to producing public service messages for groups in their communities, and will do so for little or no cost if your cause fits their criteria and they don't consider it "too controversial." If the agency itself isn't willing to help, agency personnel may be willing to help out on their own. This is a fairly standard practice in the advertising industry, since many agency writers, artists and producers are sensitive creative types who feel a little guilty about using their talents just to sell soap. (Don't confront them with this, though; it'll only make them feel defensive. Simply ask if they're interested in lending their expertise to your project.)

NATIONALLY PRODUCED PSAs

Public Media Center is a non-profit, tax-exempt media resource center. We regularly produce high quality radio, television and print advertisements on important current issues, both local and national. Our PSAs are available to broadcasters free of charge since stations will not pay for public service spots. PSAs are also available to public interest organizations at cost. Whenever possible, we ask the sponsoring organization to allow other, local organizations to put their own name, address and "tag line" message on the ads. This helps them get on the air, since broadcasters are required to serve their local community and prefer to do so, rather than serving some vague national constituency. It also makes the message more effective, since people will see it as coming not from some centralized national authority, but from citizens in their own community. And it enables the communication to be translated into action when people who hear the spots contact your organization. The process of placing "local tags" on TV spots can be somewhat expensive, however, so it isn't crucial to an effective national effort.

A national "clearinghouse" for information can direct requests to the relevant local organizations.

To find out what public interest advertising campaigns are currently available from Public Media Center, just write us or give us a call. If you're interested in having us produce materials specifically for you, please send us a brief description of what you'd like to say, who you would like to say it to, and what kind of resources (staff, informational, financial and production) you have to say it with. We'll send you our suggestions on what can be done and an estimate of how much we think it will cost to do it. If you'd like to have one of our national campaigns adapted for your own use, tell us what you want and we'll try to help you do it.

A broadcaster may object to airing your Public Service Announcement and you should be prepared to reply to the objection. In this section Media Access Project presents three objections used by broadcasters who do not want to air a PSA. MAP also suggests replies you can make to these objections.

THE THREE MOST OFTEN VOICED OBJECTIONS . . .

When you approach a station with your PSA, you will most likely encounter one of three objections. Each of these has been covered previously. We present them here in order to give you a concrete idea of what to expect.

1. *The "Controversiality Argument":*

BROADCASTER: "It is our station's position that public service announcements are a generally inadequate form for the presentation of issues, particularly when they are controversial."

A REPLY: The Federal Communications Commission defines a "public service announcement" as "...any announcement for which no charge is made and which pro--motes the programs and activities of non--profit organizations and other announcements regarded as serving the community interest."
This definition clearly states the dual purpose of public service announcements. First, to provide an opportunity for nonprofit organizations to broadcast a message of their own choosing in a format of their own choosing. Second, to serve the community interests.

Notice that nothing is said about "controversy". The dictionary defines "controversy" as "A dispute involving sides holding differing views, especially a public dispute." Any discussion of meaningful community interests will involve at least some public dispute.

(Obviously, what is termed "controversial" is itself a "controversial question." Ask the station's public service director for a list of PSAs the station has aired recently. It is easy to point to some of them as being at odds with the views of at least one reputable community organization. One station, for instance, told us that they do not accept "controversial" PSAs and then pointed with pride to a PSA aired for a gay crisis-counseling center. It is clear that homosexuality is still a "controversial issue.")

The Communications Code *requires* that you cover controversial issues of public importance. If the issue we raise is not to be covered in a PSA, then when and where are you covering it in your overall programming? Please be specific.

2. *The "Complex Public Issue" Argument:*

BROADCASTER: "It is our station's position that public service announcements are a generally inadequate form for the presentation of issues such as these. We think the public is better served by the treatment of such issues in program length form."

A REPLY: PSAs and longer public affairs programming serve two different purposes. Public service announcements interrupt daily programming with "nuggets" of information. This information is not designed to completely inform but to expose people to ideas they didn't previously know existed. PSAs remind people of something they take for granted. They arouse curiosity, provoke thought and stimulate participation. It is this exposure, intruding into regular programming, which helps to insure that someone will tune in to the longer public affairs programming when it is presented. To condemn a 30 second PSA because it doesn't inform as thoroughly as a 30 minute program is like saying arms are not good because they don't allow people to walk.

3. The "Equal Time" Argument:

BROADCASTER: "It is our station's position that 'controversial messages' will force us to give more free air-time to groups with an opposing point of view. Broadcasters are required to do this under the 'Equal Time' provisions of the Federal Communications Code. And we can't afford to give away more free air-time than we already do."

A REPLY: There is generally much confusion regarding the exact provisions of the Federal Communications Code. The rule which applies in this instance is the 'Fairness Doctrine', and *not* the 'Equal Time' rule. The 'Equal Time' provision refers *only* to candidates running for public office and states that time offered to one candidate must also be offered to his or her opponents.

The Fairness Doctrine, on the other hand, states that broadcasters must strive to reach a "reasonable balance" in their *overall* coverage of "controversial issues of public importance." If a station airs a PSA from your organization, it does *not* have to provide free PSA time for an opposing point of view. It only has to be sure that the issue is covered from all points of view in its overall programming: news reports, talk shows, public affairs documentaries, etc.

If an important public issue has *not* been covered and your organization raises the issue for the *first* time at the station, then the broadcaster *does* have an obligation to provide opportunities for an opposing point of view to be expressed. This can be done either in a PSA, or in news reports or a talk show appearance. Sure, a station can try to ignore important community issues by concentrating solely on selling more advertising. But it does so at the risk of losing its license. Broadcasters are *required* to cover controversial issues of public importance.

See the 1974 FAIRNESS REPORT of the Federal Communications Commission. Federal Register Vol. 39, no. 139 7/18/74.

If you have been denied access to air time you may want to write a letter to the station in which you state your objections to the station's refusal to air your PSA. Here is some advice from Media Access Project on how to write a convincing letter.

WRITING A LETTER

Suppose you've gone through all the channels and you've still been refused access. The next step is to write a letter to the person at the station whom you've been dealing with. What good will this do? Plenty. The key here is to send copies to this person's boss, the station manager, the station's law firm and to members of the community who have worked with you on your issue. Oftentimes you'll get the brush-off from a Public Affairs Director who feels too busy to handle your request. It may have sounded unorthodox, and he or she wants to avoid any possible trouble. Your letter shows that, if the station wants to avoid trouble, they're not going to do it by ignoring you.

Try to get these points into your letter:

1. Describe your organization. Stress your broad support in the community. Mention any prominent people who are members of your organization or who serve on your board of directors or advisory committee.
2. Describe and document the issue you are addressing. Stress the importance of the topic to your community and the timeliness of the message, by including news clippings that show your concern to be a controversial issue of public concern.
3. Try to discover any programming the station has presented on your issue. If there's been little or none, point that out. Say that, as a regular viewer of the station, you haven't seen any locally originated programming on that issue.
4. Show the station that you know what the Fairness Doctrine is. You know the station must not only balance its presentation of issues, but has an affirmative obligation to cover controversial issues in the first place.
5. Take the definition of a Public Service Announcement from FCC Form 303 and present it verbatim. Stress that the PSA *is* a proper format for the discussion of issues by members of the community.
6. Talk about the process of negotiation with the station. Has the station been fair? Reasonable? Have they *ever* lied to you? Remember that one of the few things the FCC especially frowns on is a station's dishonesty to a representative of the community--but only if you've been honest, fair and reasonable yourself.
7. If the station offered you an appearance on a Sunday morning talk show instead of an opportunity to do a PSA, then point out that one is not necessarily a substitute for the other. Sometimes broadcasters refuse you access with a vague promise that they'll produce a show for you "sometime in the future." You can grow old waiting.
8. If other radio or television stations in town are airing your PSA, then mention this!
9. In closing, discuss the reasonableness of your request. Stress the good faith that you've shown.

10. Say that you don't want to write the FCC unless you have to. But if you can't understand the station's position you'll be forced to seek relief elsewhere.

11. Ask for a meeting with the station manager to discuss your problem. Say that you will call in a specified number of days to get the station's answer. And don't forget to send copies to the station personnel mentioned above, indicating that you have done so with a "cc:" followed by each person's name and title at the bottom of the letter. If it seems appropriate to write the FCC at this point, note that in your "cc:" too, and send a copy to:

Mr. William B. Ray
Chief, Complaints and Compliance Division
Broadcast Bureau
Federal Communications Commission
1919 M Street, NW, Room 332
Washington, D.C. 20554

Mr. Ray's phone number, should a more direct approach seem appropriate, is: (202) 632-7000, ext. 26968.

Here is a model letter which could be used by grass roots groups to request time for anti-nuclear ads to counter the pro-nuke blitz. The letter was composed by Media Access Project.

MODEL LETTER TO BROADCAST STATION REQUESTING TIME FOR ANTI-NUCLEAR POWER ADS TO COUNTER PRO-NUCLEAR ADVERTISING CAMPAIGN

Dear [Broadcaster]:

I am writing on behalf of [name of organization (s) with footnote giving brief explanation of nature and purpose of organization (s)], concerning your broadcast, throughout the day and night, of several [name of company] ads that present but one viewpoint on the controversial issue of public importance concerning immediate commercial construction of nuclear power plants and use of nuclear power. These ads claim that nuclear power is environmentally clean, safe, practical, and economical.* Numerous residents of [state or community], including members of these organizations, who have been regular listeners to your station over a lengthy period of time, have informed us that they have heard no other programming on your station which presents any opposing viewpoints on this issue.

Several clippings from the [local newspaper] attest to the fact that the question of immediate construction of nuclear power plants and use of nuclear power is a controversial issue of public importance in [community], [state] and in the nation.** [If there is a pending state referendum on the issue of construction and use of nuclear power plants, or if residents are being asked to sign initiative petitions in favor of holding a referendum, this should be discussed and related to current pro--nuclear power advertising, as further documentation of the existence of a controversial issue of public importance.]

While it is true that no one group or individual has an absolute right of access to radio airwaves, a licensee must, under the fairness doctrine, provide a reasonable opportunity in its overall programming for the presentation of contrasting viewpoints on controversial issues of public importance. Moreover, the Federal Communications Commission has continuously recognized that various factors must be considered under the fairness doctrine in determining what constitutes a reasonable opportunity for response when a licensee has presented programming expressing only one side of a controversial issue of public importance. These "signposts" of reasonableness include not only the total amount of time afforded to each side, but also the frequency with which each side is presented, the size of the listening audience during the various broadcasts, the time period over which the one--sided broadcasts have appeared, and the reaching of different audiences.

In light of your broadcast of the [name of company] nuclear power ads throughout the day and night, including during periods of maximum listening, and in light of your failure to provide any adequate access for the presentation of contrasting views, according to frequent regular listeners of your station, you are invited to immediately air such contrasting views. While we are financially unable to purchase air time for such presentation, we will be happy to provide you with pre-recorded material, prepared on behalf of nuclear power opponents, which we feel would help you satisfy your fairness doctrine obligations if aired a sufficient number of times.

If you feel you have met your fairness obligation in some manner we have been unable to determine, please inform us of the specific times and substance of any contrasting programming which you have aired. Otherwise, in order to avoid our filing a formal complaint with the Federal Communications Commission, let us know exactly how you plan to meet your obligation, including whether you wish to receive our pre--recorded spots. We view this matter as extremely serious and are confident that working together we can insure a fully-informed citizenry on the issue of nuclear power.

Sincerely, yours,

---

*Copies of the text of such ads should be attached where possible.
**Copies of some of these clippings should be attached.

When you are dealing with a broadcaster whom you feel is not giving your group fair access to air time it may be necessary to use the station's public file. The public file is a set of documents and letters that the FCC requires the station to keep and, according to Media Access Project, "can be the citizen's greatest ally."

Public File

You have the right to walk into any radio or television station in the country, during regular business hours, and request to see its "public file." The station cannot ask you why you want to see it or what you want to do with the information or what organization you represent. They can only ask you your name and address.

What is the public file? It is a set of documents and letters that the FCC requires the station to keep at its main studio or at another easily accessible location in the community. The public file can be the citizen's greatest ally. This is for two reasons. First, it is a rich source of information about station practice and programming. For example, the public file must include at least the following:

-- Any pending *License Renewal Application* as well as the applications for the previous two renewal periods. (Example: If a station must file a renewal application in 1977, the public file must contain the 1977 application as well as the applications for 1974 and 1971.)

-- *Annual Programming Reports* (Form 303-A) for the past seven years.

-- *Annual Employment Reports* (Form 395) for the past seven years.

-- Results of the station's ongoing Ascertainment of the Community, including both summaries of discussions with community leaders and a general public survey about problems in the community, and census-type data on the community the station serves.

-- *Annual Problems-Programs List*, containing a list of ten community problems the station has unearthed through ascertainment *and* a list of programming the station has done over the previous 12 months to deal with those problems.

--Every letter the station has received from the public in the last three years (separated into 2 files--one on programming and one on other matters).

--A record of all commercial or program appearances and requests for time by political candidates during the previous two years.

--Composite Week Logs for the previous three years. Each log provides a detailed list of all programming and commercials by the station during a random, sample week picked by the FCC.

-- Other documents, including Ownership Reports, applications to construct new stations or facilities and FCC correspondence relating to the applications.

-- A copy of the FCC's "The Public and Broadcasting Procedural Manual."

Every bit as important as what the files contain is the question: Did the station give you free access to them? This is the other reason the public file is the citizen's greatest ally-- it is a litmus test for determining how well the station is meeting FCC rules. Therefore, you should keep a careful log of all your contacts with the station about the public file: who you spoke to; what questions they asked you; where you viewed the file; any harrassment or delay you endured; and a list of every item contained in--or missing from--the public file.

And remember: You can see the public file anytime during business hours--the station cannot ask you to make an appointment. Nor can the station personnel harrass you while you view it. For example, they can't hover over your shoulder while you go through it. They can't question you as to your motive. They can't tell you to come back another day or let you see only the parts of it they choose. You also have the right to have portions of it photocopied at a reasonable cost and within a reasonable time (less than 7 days).

NOTE: TV and radio stations must also keep on file for two years documents called the Program Logs. This is a record of every broadcast. It includes the name of every program and local commercial put on the air, the time and length of the broadcast, the sponsor of the commercial, the source of the program (e.g., network, local) and other details. You have the right to inspect and have copied the program logs but it is a more limited right than your right to view the public file. *You must make an appointment with the station and you can be asked to identify the organization you represent (if any) as well as your general purpose in examining the logs.* The station need not, in any case, make the logs available until 45 days after broadcast.

Before your group resorts to using the Fairness Doctrine to resolve your complaint with a station you should exhaust all other channels. The following information from Media Access Project explains some of the steps you should take before using The Fairness Doctrine.

APPROACHING YOUR LOCAL BROADCASTER

The purpose of the Fairness Doctrine is to ensure that radio and television stations present contrasting views on controversial issues of public importance. It is difficult, however, to convince the Federal Communications Commission that: 1) the issue with which you are concerned is, indeed, a *controversial* issue of *public importance*: and 2) there is not a fair relationship between the amount of air time given one side of the controversy and the amount of air time (if any) given your side of the controversy. In addition, litigating before the FCC can often be time-consuming and costly.

Therefore, you should always try to resolve your complaint with the station before you take your case to the FCC. (In any case, the FCC will insist that you convey your complaint to the station.) Approaching a television or radio station with a complaint is a bit of an art. Your approach (e.g., hard sell, soft sell) may vary depending on the personality of the station and your perception of its willingness to broadcast your side of the issue. In general, however, you should do the following:

1. Convey your complaint to the station by letter. Address it to the station management. It is very important that you put your complaint in writing. Otherwise, the station may later say that management was unaware of your complaint or unaware of the details.

2. Your letter should include the following:

Who you are
Let the station know that you are a *bonafide spokes-person* for the point of view which you are trying to get on the air.

What the controversial issue of public importance is.
You should identify the controversial issue with which you are concerned (e.g., Should the State of California stop the development of nuclear power plants? Should a particular dam project be continued?) Also make sure you tell the station that the issue is not only controversial but important. You might say, for example, that there have been many newspaper stories on the issue; or that the issue is part of a referendum and will be coming before the voters.

What programming the station has done which has told the "other side's" story.
It is of great importance that you make it clear to the station that you have listened or viewed program-ming which has presented the "other side" of the issue. You should identify the date and time and nature of the programming if you can (e.g., 5 spot advertisements per day for a two-week period beginning May 5, 1977, to May 19, 1977, with broadcasts at 10, 12, 2 P.M., 4 P.M., and 6 P.M.). Give as much detail about it as you can. Without this information, the station may tell you it has done little or no programming on the "other side" and tell you to go fly a kite.

The fact that you are aware of little or no programming which the station has done of your side of the issue.
Find someone in your organization who regularly listens to the station (doesn't have to be 24 hours a day). Assuming there has been no programming on your side of the issue, inform the station that you are not aware of any programming they have done to comply with their responsibility to present both sides of a controversial issue of public importance.

Request Compliance
Ask the station to comply with the Fairness Doctrine by presenting programming on your side of the controversial issue. Tell the station that you are a bonafide spokesperson for that point of view and will be glad to provide the station with your expertise and assistance. If there is a critical decision or vote approaching, inform the station of the urgency of the matter and request that it respond to your letter by a certain date.

3. Negotiate with the station. Hopefully, the station will now want to talk to you about putting on your side of the issue. The station may seek to put you on a late night or early morning half-hour public affairs program. This may be acceptable if the other side was also presented in the same format. You should strive, however, to get your message across to the public during prime time (TV) or drive time (radio). Spot announcements are probably one of the best ways to get your message across--particularly when the other side has purchased a massive amount of advertising.

This is a far cry from a complete description of the use of the Fairness Doctrine. Its only purpose is to remind you that success may often be found through initial negotiations with the station. On the other hand, if the station refuses to negotiate or offers you a good deal less than half a loaf, you may consider it in your best interest to go to the FCC anyway. Of course, the station will then go to the FCC and tell the Commission that it offered you time and you refused. So don't refuse an offer of time unless it is patently unreasonable. If it is unreasonable, but you feel you have to accept it anyway (because of the urgency of the situation), you may want to send a letter to the station telling it that you are accepting the time with an express reservation of a right to file a complaint with the FCC.

And remember. The Fairness Doctrine doesn't give you a right of access. It only compels the station to discuss the issue with which you are concerned. The station may do that through news broadcasts, public affairs programming or spot announcements. It has to use real advocates of both sides but it doesn't have to use you. In general, however, if the station knows you are a bonafide representative of the point of view which it wants to broadcast, it will take advantage of your availability.

HOW TO FASCINATE THE PRESS

By Mina Hamilton

In "How To Fascinate The Press," Mina Hamilton, who is co-chairperson of the Sierra Club's Waste Campaign in Buffalo, New York, gives some very practical suggestions for working with news reporters, holding press conferences, and getting your message across via the radio. This article first appeared in Not Man Apart *which is published by Friends of the Earth in San Francisco, California.*

The press is the lifeblood of every environmental battle. Unless citizens can obtain fair, accurate, frequent, and in-depth coverage of an issue, nine times out of ten the battle will be lost. One of the prime reasons why the Tocks Island Dam battle was not lost was because dam opponents were extremely adept at obtaining coverage in every newspaper in the region including *The New York Times, The Washington Post, The Philadelphia Inquirer,* as well as on the major networks.

One of the leaders of the Tocks battle, Mina Hamilton, a former reporter for Associated Press and *Newsweek,* attributes this success to the assiduous application of the following principles.

1. Be compassionate. Reporters are notoriously underpaid, overworked, and subjected to the unremitting pressures of last-minute deadlines, late hours, and irascible editors.

2. Be supremely polite and friendly. Introduce yourself at press conferences and hearings. Get on a first name basis. Always compliment the press on stories well done—even if it is not your story. *Stay* friendly, especially with those who haven't yet covered your issue adequately or fairly. Use the personal touch to establish a relationship. Once a relationship is established, sustain it with occasional **favors.** Feed interesting leads and tips to selected press. The favor will probably be returned by a reporter leaking information to your group.

3. *Never* phone the paper's editor to bitch and complain about inaccuracies. This is the quickest way to alienate, often permanently, the press. Do, however, in a straightforward unemotional letter to the editor, set the record straight. Never blame the reporter for what may be the publisher's policy or a night editor's error. The press is not a monolith. Be aware that the reporter, the radio announcer, the publisher, the station manager, all may have different viewpoints. Don't chew out the radio announcer for a policy that he may secretly disagree with and is slowly trying to change from within. Don't bad mouth the press to friends, legislators, politicians. Word will somehow always get back to the paper and damage your reputation.

4. Be devastatingly accurate. Never make statements that cannot be substantiated. Never make unfounded allegations or personal slanders--even if it is during supposedly off-the-record personal conversations. (A safe bet with all news media is that *nothing* is off-the-record, no matter what promises are given.) Marshal the facts, carefully, logically. As often as possible, refer to credible studies, official documents, hard statistics to justify a position. Avoid the general. List specifics, detail cases. If an organization can cite the eleven historic homes due to be bulldozed, the 52 cases of land acquisition abuses, or 17 examples of misleading statements in an environmental impact statement, it makes a better story for the press and also, very importantly, makes it look as if you know what you are talking about. If, during a phone interview, you do not have the answer to a question, don't fudge it. State that you want to be sure to have the statistic correct, look it up and call the reporter back. For a live interview, however, the citizen must have all the facts at his/her fingertips, on 3x5 note cards, if not in the head.

5. Be interesting. State the facts, but do so using colorful, evocative language. Create dramatic events. Demonstrations, pickets, bus tours, float trips, map unveilings are all valuable techniques for engaging the press. At hearings, conferences, and debates, have charts, maps, and photographs available for the media. The press is tired of words, nothing but words. A good, captioned photograph will often benefit the cause far more than pages of copy. Have enough copies of the photo for all members of the press **present.** If a citizen is appearing on a local television program, suggest he/she bring along a selection of color slides. Often, TV stations like to use photos as a backdrop for interviews. A story without much drama can be made more lively for the press. An analysis of the biases in a government study, for instance, will get more coverage if the activist sends a bunch of protest

telegrams to Washington, followed by a press release about the telegrams *and* the organization's critique of the study. (Like all recommendations in this article, this technique must be used judiciously.)

Get tuned into what interests the press. Here is the Waterloo for most environmentalists. If your local media do not care about "environmental issues," don't continue to harp on the subject. Temporarily, soft pedal wilderness or wildlife. Find the angle that concerns the press. Go after the economics of the project, tax-impact, efficiency in generating jobs, energy conservation, human interest stories, political scandals, historic homes. When the press is finally interested, *then* the activitst can shift back to environmental issues.

6. Be convincing and methodical. Frequently, it is hard for a small, newly formed group to develop credibility. The group's press releases never appear or are buried on page 62. A good remedy is to join forces, temporarily, with an established organization that already has some press credibility. Several joint press releases sent out by the local group *and* a Public Interest Research Group or nationally known conservation group will help build recognition. Later, the local group can operate independently. If the new group is unfamiliar with the art of the press release, it is a good idea to contact a group in the state that is getting good press coverage and ask to see examples of their press releases.

With press releases, first appearances are half the battle. Any environmental organization which sends out smudgy mimeographed statements flawed by typos and sentence insertions deserves what it gets—oblivion. A xeroxed, clean copy, double-spaced with generous margins is mandatory. The release should be written in a lively, informative style. The first sentence must catch *and hold* the attention of a harassed, tired, bored reporter. Pithy quotes are a good technique for brightening the statement.

Keep on patiently sending out the press releases. Do not give up if they do not appear. Even if a release does not get published, it is important for the press to know that you exist and that the issue is alive. If after four or five tries, the press releases still are not appearing, there is probably something wrong with *your* press releases. Consult a professional. (You may be sending out too many press releases on non-news events.)

Find out press deadlines for all the media you have contact with. Don't expect a reporter to cover a story you dump on his desk two hours before filing time.

7. Be consistent. Have one spokesperson for the organization who handles all contacts with the press. The spokesperson must be persistent, polite, extremely articulate, totally unflusterable, and massively well-informed. He/she must be skillful at avoiding lawsuits, oversimplifications, and misleading statements. A favorite reportorial interviewing technique is to preface a statement with "Wouldn't you say that. . . ." If the spokesperson answers either yes or no, the reporter's statement may end up as the spokesperson's quote!

8. Be wary of press conferences. Most groups schedule too many conferences about insignificant events. The press conferences must involve big news, big names. If it's just another slanted government study, inadequate environmental impact statement, initiation of a lawsuit, you should probably go for a press release. If a press conference is in order, be sure to contact a friendly source in the media regarding scheduling so as to avoid conflicts. Try to schedule the conference to dovetail with another event related to your issue at which the press is *already* going to be present so as to assure good coverage. Have coffee and refreshments available, plus plenty of packets of information, visual displays, and photos. *Keep the presentation short* and leave lots of time for press questions.

9. Be on the radio. Don't forget the potential for radio public service announcements—PSAs. Many stations will run PSAs *ten times per day* for several weeks at a time so the opportunity for publicity is impressive. Start monitoring your local radio stations. Note type, frequency, and quality of PSAs. Call the stations and talk to them. Describe what you would like to do, find out what is likely to be accepted by the station. Be accommodating. Most stations prefer 30-second tapes which can be dropped into a cartridge for a couple of weeks. An oblique, soft-sell approach which publicizes the name of your group in a gracious way is the best way to start. If your stations prefer PSAs with music, give it to them. The tape will be aired more frequently. One approach might be a few seconds of guitar music followed by an announcement "This music brought to you by the Delaware Valley Conservation Association which wishes to thank canoeists and campers for not littering the beautiful Delaware River and Valley." Who knows? Your next message may be: "The Delaware Valley Conservation Association wants to thank citizens for their support on keeping the Delaware River free-flowing." This last message was suggested to us *by* a radio station. In short, the media can be tamed!

BEING INTERVIEWED FOR TELEVISION OR RADIO: HOW TO PREPARE

By Citizen Involvement Training Project

The following will help activists prepare for a radio interview and evaluate the success of the interview as well. For a copy of "A Citizen's Guide to Using the Media for Social Change," by Robbie Gordon, write to CITP, c/o the University of Massachusetts at Amherst.

The first few times any layperson is interviewed on television or radio it is difficult to beat back the voice inside you that keeps saying "there are 20,000 people listening to me right now." That's why it's really helpful to spend some time preparing for the interview, and it is especially helpful for the entire organization to take part in this process.

Going through a mock interview with your group helps on many counts. First of all, it gives all of you a chance to think through all the possible questions and curves you might be thrown during the interview. It can help you think through your answers, understand how the rest of the group will react to your answers (and, you hope, head off the posse that might be waiting for you when the program's over) and will give you an idea of what the group as a whole feels the "party line" ought to be when and if certain touchy subjects arise. It will also give you an idea of the most important points you'll want to cover—the most important ideas and facts you want to get across—and can help you shape your impromptu answers to fit the reply you have in mind, even if the question might not have been aimed in that direction (this is not to advocate smoke-screening or fancy footworking, rather, to ensure your message gets across even with a poor interviewer).

Mock interviews can also prepare the rest of the group for future interviews (in case others will be going on the air at some other time) and can help you get a lot of feedback and suggestions about your personal style (too many "ums," swallowing the ends of sentences, etc.)

Debriefings after an interview can be helpful for the entire group and for future appearances. An early debriefing with members of our own project revealed that swivel chairs were something to be cautious of (our interviewees were prone to swivel constantly, and not synchronously). It also made us more aware of the need for eye contact with the interviewer (in the case of TV).

In some cases the interviewer will ask you to prepare the questions for the interview yourself and bring them to the interview. Usually that's a good warning signal that the interviewer will not be very attached to the subject in question and may not even be listening to your responses. Other interviewers don't even want to see you or talk to you before you're on the air, so they can become engaged in a lively and fresh dialogue when the times comes. Still others want at least some background information (written or verbal) on the group and the issues ahead of time so they can prepare their own questions. You should try to find out what style you're dealing with ahead of time so that you can know what to expect.

If the program is a live call-in show, chances are likely that the producer will want to book you on the program only if you have something controversial to talk about (something that will prompt many phone-ins). You may want to think it through with your group to be sure you can pique some lively reactions (but at the same time avoid combustion).

If the program is a television talk show, the producer may ask you to provide some visuals (videotape, slides, demonstrations, etc.)—something to make the segment more lively and illustrate what you're talking about (so you don't end up with a bunch of "talking heads").

HOW TO USE THE MEDIA TO PUBLICIZE ANTI-NUCLEAR TRIALS

A courtroom can be an important forum for educating the public about the dangers of nuclear power. But of course not everyone living in the area where the trial is being held will be interested in or able to attend the trial. The information they receive will come from radio, television and press coverage of the trial. This is why it is so important to develop rapport with the press, know how to make use of free television and radio time, and write effective letters to station managers. In many ways the information we make available to the public about the dangers of nuclear power is more important than whether we are found guilty or not guilty for a crime or misdemeanor. The following advice is from the Anti-Nuclear Legal Project's Pro Se Handbook.

Pro Se and the Media

The media can serve two very important purposes during your trial: First, they can help you turn the trial into an educational process for the general public; second, they can increase public awareness of your particular group. In the best of all worlds, the media can actually serve as a recruitment tool.

The steps outlined below can help you set up a comprehensive media plan. We hope that you find it helpful.

Establish a media contact list. This should be done long before the trial. List the names, addresses and telephone numbers of the media in your area that could possibly cover the trial. Local newspapers, radio stations and television stations should be listed, but don't overlook the shopping rags, group newsletters and the weekly or monthly magazines that are published in your area.

Establish personal contacts. Your group should have one or more designated press contacts who can personally call reporters, editors and newscasters to inform them of the impending trial and its significance to the community. There are a lot of ways to get your foot in the door—one way is to send out a press release announcing the location of your office, the name of your press contact, or some other created news item. Follow up the release with a phone call, asking the editor or reporter if they need further information. Get the name of your contact and keep it with the media list. This work is time consuming, but it may pay off in good press relations and in some advance coverage.

Set up a clippings file. This will give you an idea of the media's response to your group. Whenever the coverage is good be sure to thank the reporter, through a letter or phone call. Reporters are often working under tight deadlines and demanding editors—they will appreciate and remember your thank you.

Press releases. Use the sample press release as a guideline. Always include names, dates, addresses, and phone numbers of the press contact people. Many editors will rewrite your release or simply use it as reference during the trial. Others may simply print it in the form you send, so keep the writing clean and concise. Members of the media may choose to ignore your group, but don't stop sending releases because of this. If coverage is important to you, you've got to be persistent.

Your news releases should be distributed to all print, radio and television media. Be sure to learn the deadlines for all of these. All news conferences should be held in time to meet the deadline for the evening news broadcasts; remember, the 5 p.m. news may have a 1 p.m. deadline.

Press conferences. Press conferences should be scheduled as far in advance as possible. If time permits, the media contacts should call as many reporters and newscasters as possible. Try to create some "visuals" for television—large charts, graphs, and diagrams are appropriate.

Schedule a press conference for any expert witnesses who will testify at the trial. If you are bringing in more than one expert, you can hold a press briefing, where all experts are available to make statements and answer questions from the press.

Letters to the editor—Editorial replies. Respond in writing to any appropriate article. A pro-nuke article deserves to be challenged, and anti-nuke articles should be reinforced. Regardless of position, make sure you include your address and phone number in your letter to the editor.

Broadcast editorials on radio and T.V. should also get a response. Station managers will often solicit a response if they know of groups interested in a particular issue. For this reason it is wise to send a letter to the station manager, stating your position and asking for the opportunity to reply to appropriate editorials. Broadcast replies are usually limited to a few minutes at most, so choose your words wisely, and make sure you get in the group's name and telephone number.

Develop a list of talk shows on radio or television that deal with local issues. Contact the show manager and try to schedule an appearance for a member of your group. Another approach would be to schedule an appearance of a local "expert" who supports the anti-nuke movement—a scientist or professor at a university may be willing to appear and discuss nuclear hazards, etc.

Talk show hosts can be tough, so be sure to set up several practice sessions where the guest is interviewed by members of the groups asking tough questions.

Learn the public service policies of the local media. Radio and newspapers will often carry free announcements at no charge to community groups. Use this service to announce meetings, fund raising events and planning sessions.

Try for some feature coverage. Watch the newspapers, looking for feature articles that in some way relate to the nuclear issue. For example, watch for environmental articles and contact the writer, suggesting that your group may have some information that would make a good feature. If no one on the paper will cover the issue, write the feature yourself. Submit it with some clean photographs. These may help you get the feature printed.

National coverage. Your phone book will list the nearest contacts for national media. Many of the magazines have local bureaus in the larger cities throughout the country. Contact the nearest one with a good cover letter, a copy of a recent release, and some copies of recent news clips. Your cover letter should make a convincing pitch about getting coverage, so write it carefully. Try to articulate the most interesting parts of the case—pro se techniques, choice of evils defense, and non-violent or collective group aspects are a few positions to present. Follow up any letter with a phone call, asking for feedback.

For broadcast news, your local affiliated stations should be able to give you national contact names. Don't overlook the public radio network, which has excellent news coverage and often accepts news from freelancers all over the country.

HOW TO WRITE A PRESS RELEASE

By Energy Policy Information Center

Press Releases are a very important part of using the media. Boston's Energy Policy Information Center offers a few suggestions on how to write a release and provides a sample release for grass roots groups to study.

Media outlets are much more likely to use a press release if it is written in AP (Associated Press) style, the style they are used to. The release on the next page is an example of a standard release. This one was designed to be released the day the IRG report came out, and was picked up in the Wall Street Journal. (One sentence, that is.)

A release should have a lead, an opening paragraph that makes the most important point of the release. The lead should then be backed up by the rest of the release. Thus the phrase "foolishly optimistic" is found in the full quotation below.

The best style uses clean, short sentences and short paragraphs, with only one main idea in each graph. If you've got more than one idea to express, then break the ideas down.

All opinions and direct quotes should be attributed to a person. Fact and opinion must be kept strictly separated. Nothing will kill a release faster than an over-emotional mixture of fact and opinion. Try to avoid giving vent to the real anger you sometimes feel in commenting on government or industry policies. Expressions like "this stupid, ridiculous report," may be what you feel, but your release will wind up in the waste basket.

FOR IMMEDIATE RELEASE   Contact: Dick Bell

Thursday, October 19, 1978   Steve Hilgartner

WASHINGTON–The federal Interagency Review Group (IRG) report on nuclear waste management is "foolishly optimistic and a travesty of public participation," a spokesperson for the WASTE WATCH campaign of the Energy Policy Information Center (EPIC) said today.

"We need a policy for making well-informed decisions about the disposal of radioactive waste," said Dick Bell, an EPIC spokesperson. "But instead the IRG report is an attempt to bail out the floundering nuclear industry with a public relations blitz."

California, Wisconsin and Maine have all banned further nuclear construction until a waste disposal solution is found.

"But there is no such solution at hand," Bell said. "The IRG's foolishly optimistic report neglects recent geologic evidence that proves there is no **safe** disposal method available today. Recent reports by the EPA, the U.S. Geological Survey, and the Office of Science and Technology

-more-

Policy all agree that we know far too little about the earth to be making decisions about waste disposal now.

"But the IRG wants to go ahead with the Waste Isolation Pilot Project in New Mexico, and with the construction of Away-From-Reactor storage facilities all over the country.

"These are not solutions," Bell said, "they are a sleight-of-hand maneuver to conceal the fact that no solution exists.

"We've seen this public relations act before with the Rasmussen report. When the public was concerned about reactor safety, an academic hit-man was brought in from MIT, and reassuring numbers went out in industry advertising costing tens of millions of dollars a year. Only now, three years later, has the NRC admitted Rasmussen's numbers were basically worthless."

During the past 30 years, more than 5000 studies failed to come up with a solution to the waste problem. Carter ordered the IRG to come up with a policy in six months. "It's like demanding a cure for cancer by next week," Bell said.

The IRG held only three public hearings in the entire country, with only one east of the Mississippi River. "That was enough of a travesty," Bell said, "but then the IRG chose to ignore most of what it heard." At all three public hearings there were repeated calls for a moratorium on the production of additional waste until a permanent disposal method is devised.

"The IRG has categorically refused to consider calling for a moratorium as one of its possible recommendations," Bell said.

EPIC has organized WASTE WATCH, a national campaign focusing on the IRG and nuclear waste policies. WASTE WATCH distributes organizing literature on nuclear waste and will sponsor regional organizing conferences on the issue. The first WASTE WATCH conference will be held in New England in early 1979.

ATTN NEWS DIRECTORS: The Energy Policy Information Center will hold a news conference on the IRG in its Boston office, Friday, October 20, at 1:30 PM.

BIBLIOGRAPHY OF REFERENCE MATERIALS CONCERNING ACCESS TO BROADCASTING
(Selected Listing)

I. PERIODICALS

Access. Bi-weekly periodical (except the last issues of August and December, National Citizens Committee for Broadcasting, 1028 Connecticut Avenue, N.W., Washington, D.C. 20036. Subscriptions: $36 yearly (institutions); $24 (individuals); $18 (students). *Access* No. 34 (May 17, 1976) includes a special section entitled, "Whys and Hows of Public Service Announcements," pp. 8-21.

Broadcasting Magazine. Broadcasting Publications, Inc., 175 DeSales Street, N.W., Washington, D.C. 20036. Subscriptions: 1 year, $40; 2 years, $75; 3 years, $105.

Columbia Journalism Review. Bimonthly magazine published under the auspices of the faculty, alumni and friends of the Graduate School of Journalism, Columbia University. Columbia Journalism Review, Subscription Service Department, 200 Alton Place, Marion, Ohio 43302. Subscription: $12 yearly.

CPB Report . . . The Newsletter of the Corporation for Public Broadcasting. Bi-weekly newsletter. Corporation for Public Broadcasting, 1111 16th Street, N.W., Washington, D.C. 20036.

Media & Methods. (For Teachers.) Magazine published nine times during the school year by North American Publishing Company. Media & Methods, 401 North Broad Street, Philadelphia, Pennsylvania 19108. Subscriptions: one year, $11; two years, $20; three years, $28.

Media Report to Women. Monthly newsletter. Women's Institute for Freedom of the Press, 3306 Ross Place, N.W., Washington, D.C. 20008. Subscriptions: 1 year, $20; 2 years, $38.

Public Telecommunications Review. National Association of Educational Broadcasters, 1346 Connecticut Avenue, N.W., Washington, D.C. 20036. Per copy: $3.00. Subscriptions: 1 year, $18.

Radio Waves. . .The Guide to Radio in Eastern Massachusetts. Monthly Magazine. Radio Waves, Inc., 101 Tremont Street, Boston, Massachusetts 02108. Subscriptions: 1 year, $12 (12 issues).

Televisions. Quarterly magazine. Washington Community Video Center, Inc., P.O. Box 21068, Washington, D.C. 20009. Subscriptions: 1 year, $10 (pre-paid); $15 (if billed); $15 (institutions).

TV Guide. Weekly magazine published by Triangle Publications, Inc., TV Guide, Box 400, Radnor, Pennsylvania 19088. Subscriptions: 27 weeks, $7.95; 47 weeks, $13.95; 52 weeks, $15.50.

II. HANDBOOKS AND PAMPHLETS

Getting in Ink and On the Air (1973). Publicity handbook. Metropolitan Cultural Alliance, Inc., 250 Boylston Street, Boston, Massachusetts 02116. (Currently out of print: call the Alliance at 247-1460 for further information).

Getting Into Print and Breaking Into Broadcasting (1978). Pamphlets. The League of Women Voters, 1730 M Street, N.W., Washington, D.C. 20036. (Publication Nos. 484 and 586, respectively.) Per copy: $.25.

If You Want Air Time. . .A Publicity Handbook (1978). Pamphlet. The Station Services Department, National Association of Broadcasters, 1771 N Street, N.W., Washington, D.C. 20036.

Media Access Guide. Handbook. The Boston Community Media Council, Inc., 566 Columbus Avenue, Boston, Massachusetts 02118. Per copy: $1.

Media Action Guide (1979). Pamphlet. Massachusetts Social and Economic Opportunity Council, 294 Washington Street, Boston, Massachusetts 02108. Free.

Media Action Handbook. The National Committee Against Discrimination in Housing, 1425 H Street, N.W., Washington, D.C. 20005. Per copy: $3.

The Media: How to Use It Effectively (November 1978). A Layman's Guide for Volunteer Organizations. The New England Environmental Network and the Lincoln Filene Center for Citizenship and Public Affairs, Tufts University, Medford, Massachusetts 02155. 1-50 copies: $2/each. Reduced rates for more than 50 copies. Tel. (617)-628-5000 x 508.

Strategies for Access to Public Service Advertising (1976). Handbook. The Public Media Center, 2751 Hyde Street, San Francisco, California 94109. Tel. (415) 885-0200.

We Interrupt This Program. . .A Citizen's Guide to Using the Media for Social Change. Citizen Involvement Training Project, 138 Hasbrouck, U/Mass, Amherst, Massachusetts 01003. Per copy: $5; 5-19 copies, 10% discount; 20+ copies, 20% discount.

III. BOOKS

Barnouw, Erik. *Tube of Plenty. . .The Evolution of American Television.* Oxford University Press, New York, New York. 1975.

Barron, Jerome A. *Freedom of the Press For Whom?. . .The Right of Access to the Mass Media.* Indiana University, Bloomington, Indiana. 1973.

Brown, Lee. *Television. . .The Business Behind the Box.* Harcourt Brace Jovanovich, Inc., New York, New York. 1971.

Cater, Douglass, editor. *Television as a Social Force.* Project editor: Richard Adler. Sponsored by the Aspen Institute on Communications and Society. Praeger Publishers, New York, New York. 1975

Diamond, Edwin. *The Tin Kazoo.* MIT Press, Cambridge, Massachusetts. 1975

Epstein, Edward Jay. *News from Nowhere . . . Television and the News.* Random House, New York, New York, 1973.

Friendly, Fred W. *The Good Guys, The Bad Guys and the First Admendment . . . Free Speech vs. Fairness in Broadcasting.* Random House, New York, New York. 1975

Powers, Ron. *The Newscasters.* St. Martin's Press, New York, New York. 1977

Schmidt, Benno C., Jr. *Freedom of the Press vs. Public Access.* Sponsored by the Aspen Institute Program on Communications and Society and the National News Council. Praeger Publishers, 111 Fourth Avenue, New York, New York 10003. 1976.

Shanks, Bob. *The Cool Fire . . . How to Make It in Television.* Vintage Books. Random House, New York, New York. 1976.

IV. REPORTS

Broadcast Reform at the Crossroads (June, 1978). By Anne W. Branscomb and Maria Savage. Kalba Bowen Associates, Inc., 12 Arrow Street, Cambridge, Mass. 02138. Per copy: $15.

A Public Trust . . . The Landmark Report of the Carnegie Commission on the Future of Public Broadcasting. Carnegie II Report. Bantam Books, New York, New York. 1979.

V. DIRECTORIES

Broadcasting Yearbook. (Annual) For information, write: Broadcasting Publications, 1738 DeSales Street, N.W., Washington, D.C. 20036.

Citizens Media Directory. (April, 1977). National Citizens Committee for Broadcasting, 1028 Connecticut Avenue, Washington, D.C. 20036. In process of being revised and reprinted: please contact NCCB at 202-466-8407 for further information.

Editor and Publisher Yearbook. (Annual) For information write: Television Digest, Inc., 1836 Jefferson Place, N.W., Washington, D.C. 20036.

1978 Guide to Film and Video Resources in New England.. Film Information Office, University Film Study Center, Inc., Box 275, Cambridge, Massachusetts 02138. Per copy: $5 plus .50 postage.

1978 NAEB Directory of Public Telecommunications. For information write: National Association of Educational Broadcasters, 1346 Connecticut Avenue, N.W., Washington, D.C. 20036.

New England Media Directory. (Annual) New England Newsclip Agency, Inc., 5 Auburn Street, Framingham, Massachusetts 01701. Per copy: $18 plus .50 tax.

Television Factbook. (Annual) For information write: Television Digest, Inc., 1836 Jefferson Place, Washington, D.C. 20036.

HOW TO OPPOSE NUCLEAR POWER IN THE CLASSROOM

Having survived the crises of Three Mile Island, the nuclear power industry is mounting a new and more fervent campaign to convince the American people that atomic energy is vital to our nation's welfare. One of the primary targets of the industry's propaganda will undoubtedly be the schools. What can a teacher do to counter the pro-nuke blitz? How can we challenge full page ads in Time *which assert that Three Mile Island demonstrated nuclear power is safe? And how can we show our students that nuclear power is unnecessary and dangerous?*

ADVICE TO TEACHERS

By Fred Wilcox

In this section I have drawn upon my experience as a teacher to provide a number of techniques which can be used to educate students about the dangers of nuclear power.

In order to be an effective opponent of nuclear power, whether in or out of the classroom, we must first learn how reactors work; why the United States has spent over 17 billion dollars developing what the *Wall Street Journal* calls *Atomic Lemons*; who really benefits from nuclear power; why nuclear waste is so dangerous (why it can neither by stored nor transported safely); how the development of nuclear power threatens civil liberties; and much more. Although this may seem an overwhelming task at first, there are several excellent books on nuclear power which provide this essential information. I will mention just a few.

The book I would recommend teachers read first is *No Nukes, Everybody's Guide To Nuclear Power* by Anna Gyorgy and Friends. Written in a clear and easy to understand style, *No Nukes* describes the various phases of the nuclear fuel cycle, examines the dangers which each cycle poses to the environment and human health, and challenges the idea that nuclear power is economical, clean, or safe. *No Nukes* is a large book, but its size belies the ease with which it can be read and understood. I would suggest that two or three teachers share a copy of *No Nukes*.

Another excellent book is John Gofman's *"Irrevy," An Irreverent, Illustrated View of Nuclear Power,* a collection of talks given by Dr. Gofman at anti-nuclear demonstrations. Dr. Gofman is one of the best informed and most outspoken opponents of nuclear power. He believes the utilities are "committing premeditated random murder" when they construct nuclear plants and offers many statistics to prove that nuclear power is dangerous. *Irrevy* can be purchased at most book stores or acquired by writing: Committee For Nuclear Responsibility, 2140 Taylor (room 1101), San Francisco, Ca. 94133.

For an explanation of how radiation affects the human body, particularly women and children, Helen Caldicott's *Nuclear Madness* is very good. Dr. Caldicott is a pediatrician at Childrens Hospital in Boston and has seen many children die from leukemia. She writes with passion and anger, calling nuclear power the "ultimate medical insanity." *Nuclear Madness* can be purchased at most book stores.

The Anti -Nuclear Handbook by Stephen Croall and Kaianders is a Pantheon Documentary Comic Book which depicts the development of nuclear power and its dangers in a lively and sometimes amusing way. The handbook is an excellent way of making technical material interesting to students who might not be willing to read *No Nukes* or other books about nuclear power.

Dome by Lawrence Huff is a fictional account of an accident at a breeder reactor and could be used by English teachers to convey the dangers of nuclear power to fiction buffs.

Reading these five books will not make anyone an expert in the field of nuclear power, but they do provide basic background information enabling a teacher to counter many of the absurd arguments the utilities and their government spokespeople make for this deadly source of electricity. Each of these books can be comprehended by the average ninth grader.

There are many excellent films about nuclear power and the filmmaker/producers are often willing to negotiate fees if you are unable to pay the full rental rate. A very effective visual aide is Helen Caldicott's *I Have Three Children of My Own,* a sixteen minute slideshow in which Dr. Caldicott discusses the dangers radiation and plutonium pose to human health. The Trumansburg Rural Alliance has shown this slide show to religious and civic groups whose members ranged in age from five to sixty-five, and the discussions which followed have always been lively and very informative. *I Have Three Children of My Own* is a scathing indictment of an industry responsible for making plutonium, which is causing increasing numbers of humans to die from cancer, mutating our genes so that our offspring will be born with "little feet sticking out of their shoulders or little hands sticking out of their sides," and poisoning the biosphere for two hundred and fifty thousand years. This slide show can be shown to all grade levels, from kindergarten to high school seniors, and can easily be set up in the teachers' lounge for viewing during lunch break or before teachers leave for the day. *I Have Three Children of My Own* can be obtained from:

Packard—Manse Media Project
P.O. Box 450
Stroughton, Mass 02072

Nuclear Information & Resource Service has compiled a list of films which can be used by educators. The list includes films about citizen action against nuclear power, the fission process, and the exposure of civilians to radiation from government bomb tests in the Nevada Desert.

FILM LIST

The following is a list of films which are available from commercial distributors. Videotapes of the films may also be available. Rental costs may vary considerably from listed prices; grass roots persistence may gain lower-cost or even "free" rental agreements. Handling and mailing costs vary widely. Finally, the major distributors listed may not be the only or even the most convenient place to get copies of the films. Check your local library, college a-v department, state energy office, neighboring safe energy group or even the electric utility (which is sure to have a film library stocked with industry recommended titles). Many of these groups can be convinced to purchase or rent the films in the public interest.

Bullfrog Films, Inc.
Oley, Pennsylvania 19547
(215) 779-8226

No Act of God
Experts Alvin Weinberg, Theodore Taylor, Hannes Alfven, and Amory Lovins debate.

38 min., color, 16mm
Sale: $695
Rental: $35

Campaign for Political Rights
201 Massachusetts Ave. NE
Washington, D.C. 20002
(202) 547-4705

The Intelligence Network
A general introduction to the government's use of police powers to suppress lawful activity, including anti-nuclear activities.

35 min., color, 16mm
Sale : $350
Rental: $45

Churchill Films
662 Robertson Blvd.
Los Angeles, CA 90069
(231) 657-5110

Energy: The Nuclear Alternative
Explores what fission power is and how it works, as well as

controversies about the nuclear fuel cycle and radioactive waste.
20 min., b/w
Sale: $290
Rental: $21

Films, Inc.
1144 Wilmette Ave.
Wilmette, Illinois 60091
(800) 323-4222 / (312) 256-4730

Danger: Radioactive Wastes
Network television's treatment of the nuclear waste issue.

50 min., color, 16mm
Sale: $575
Rental: $40

Green Mountain Post Films
P.O. Box 177
Montague, Massachusetts 01351
(413) 863-4754

Better Active Today Than Radioactive Tomorrow
Documents the story of the first mass occupation of a nuclear plant site, in Whyl, West Germany. Actions by local citizens and farmers are seen through the eyes of a local film-maker.
65 min., b/w, 16mm
Sale: $700
Rental: $40, $55, $80

The Last Resort
Examines the Seabrook, N.H. battle over the construction of a large nuclear station, and the roots of citizen opposition intending to thwart the utility's plans.
60 min., color, 16mm
Sale: $625
Rental: $35, $50, $75

Lovejoy's Nuclear War
Tells the story of Sam Lovejoy, who toppled a 500 foot nuclear weather tower to symbolize his opposition to the planned Montague, Massachusetts reactors, and of his subsequent trial.
60 min., color, 16mm
Sale: $625
Rental: $35, $50, $75

More Nuclear Power Stations is a Danish documentary examining, in a cool and detached manner, the inner workings of nuclear power plants, from reactor core to waste disposal.
55 min., color, 16mm
Sale: $625
Rental: $35, $50, $75

Radiation and Health summarizes a lively debate on the health effects of radiation between Dr. Helen Caldicott, a pediatrician and anti-nuclear activist, and Reginald Rogers, a utility spokesman.
15 min., b/w, 16mm
Sale: $125
Rental: $15, $20, $25

Sentenced to Success
Produced by a French atomic workers' union, the film documents the operation and risks of nuclear reprocessing at the
giant facility in La Hague, France.

60 min., color, 16mm
Sale: $625
rental: $35, $50, $75

Training for Non-Violence
Using a 1976 and 1977 Seabrook footage, the film reveals the
preparation and practice of non-violent civil disobedience.
Civil rights and peace activists comment.

20 min., b/w, 16mm
Sale: 625
Rental: $15, $20, $25

National Training Fund
1900 L St. NW, Suite 405
Washington, DC 20036
(202) 833-9543

Under the Sun
is a pro-solar film prepared for viewing by organized labor.

26 min., color, 16mm
Sale: $150
Rental: free

New Mexico People & Energy Research Project
810 Vassar NE
Albuquerque, New Mexico 87106
(505) 266-5009

People and Energy in the Southwest
Describes the effects of nuclear development on the Navajo
Nation in New Mexico, which supplied the majority of urani—
um miners to nuclear fuel companies. The miners suffer very
high rates of cancer.

40 min., 140 color slides & script
Sale $125 (individuals and citizen groups); $200 (institutions)

New Time Films
1501 Broadway Suite 1904
New York, New York 10036
(212) 921-7020

Paul Jacobs & The Nuclear Gang
Covers exposure to atomic fallout by civilians around the gov--
ernment's Nevada nuclear test site in the 1950s and early 1960s;
soliders who were participants in the nuclear bomb tests; work—
ers at the government's nuclear facilities in Hanford, WA and
Portsmouth, NH; and the civilian population who live down—
wind from the Rocky Flats, Colorado plant.

60 min., color, 16mm
Sale: $750
Rental: $75

McGraw-Hill Films
1221 Avenue of the Americas
New York, New York 10020
(212) 997-2813

Nuclear Power: Pro and Con
Presents both sides of the nuclear question by addressing
issues such as safety regulations, efficiency, and cost.

50 min., color, 16mm
Sale: $725
Rental: $40

Public Television library
Public Broadcasting System
475 L'Enfant Plaza SW
Washington, DC 20024
(202) 488-5220

*The Advocates: Should the United States Expand its Nuclear
Power Program?*
In a trial-like discussion of nuclear energy. Anthony Roisman,
David Comey, and John Holdren argue against nuclear; Charles
Wilson, Richard Wilson, and Rep. John B. Anderson argue for
fission. Former Governor Michale Dukakis of Massachusetts
moderates.

60 min., color, 3/4" video cassette
Sale: $230
Rental: $109.25

*The Advocates: Should We Stop Construction of Nuclear Pow-
er Plants?*
Anthony Roisman, Michio Kaku, Charles Komanoff and Barry
Commoner say yes; Avi Nelson, A. David Rossin, and Alan
Manne say no. Governor Dukakis of Massachusetts modera-
tes.

60 min., color, 3/4" video cassette
Sale: $230
Rental $109.25

Plutonium: An Element of Risk
Before Jack Lemmon did the *China Syndrome,* he narrated
this investigation of proposals to expand the use of plutonium
as nuclear fuel. A controversial program, many stations refused to air this story.

59 min., color, 3/4" video cassette
Sale: $200
Rental: $95

Time-Life Films
100 Eisenhower Drive
P.O. Box 644
Paramus, New Jersey 07652
(201) 843-4545

Incident at Brown's Ferry
Public television's *Nova* series examines a seven hour fire
with ominous implications at the world's largest nuclear power plant in Alabama. Started by a candle, the fire caused a
50 month shutdown and $150 million in damages.

58 min., color, 16mm
Sale: $650
Rental: $ 60

The Nuclear Dilemma
By developing nuclear fission plants, "Are we creating a doomsday machine?"

40 min., color, 16 mm
Sale: $450
Rental: $45

The Plutonium Connection
Robert Redford narrates this study of the public threat posed by plutonium, a key weapons ingredient.

59 min., color, 16mm
Sale: $590
Rental: $60

The Pomeroy File
The story of a Texas anti-nuclear activist who becomes the subject of the Texas Public Safety Department, uncovering a web of private, state, and corporate spy forces.

17 min., color, 16 mm
Sale: $215
Rental: $25

FIELD TRIPS

One way to demonstrate to your students that the nuclear industry is more interested in public relations than the health and welfare of the American people is to take your classes on a field trip to a nuclear plant's information center. Before you visit the center you should spend a few weeks studying and discussing nuclear power, showing films about the effects of radiation on human beings and making certain your students will challenge the tour guide with intelligent questions.

It should be kept in mind that information centers are located near nuclear plants and, although it is highly doubtful that spending an hour inside the center will endanger you or your students, you might consider calling the plant before your visit to make certain it has not scheduled a *routine* release of radioactivity on the day of your visit.

Here are a number of questions you might ask:

Why the plant you are visiting remains open when there is no reliable way to safely store the waste it is producing.

▶ How often the reactor is shut down for repairs. (Most reactors in the United States are shut down fifty percent of the time.)

▶ Who manufactured the reactor; whether it is similar to the one that failed at Three Mile Island, and whether the safety backup systems have actually been tested.

▶ Ask to see a copy of the plant's evacuation plans for the surrounding area in the event of an accident. If the plant does not have an evacuation plan, demand to know what steps are being taken to develop a plan.

▶ Ask about intentional and accidental releases of radiation from the plant and whether the tour guide understands that radioactive isotopes can enter the human body and cause cancer.

▶ Ask about the condition of the spent fuel pool and which company the plant has hired to remove the high level waste to a storage site. (Before you visit the information center you might want to study the accident records of the waste transporting companies. See "Transportation Ban" section.

▶ Ask who inspects the packaged high level waste *before* it leaves the plant, how transportation routes are decided, and what kind of notification is given to communities through which the waste will be traveling.

▶ Ask to see documentation of the plant's accident record, whether the plant has suffered any breakdowns due to leaking or ruptured pipes, computer malfunction, fires or human error.

The tour guide may be young, uninformed or misinformed, and "only doing his/her job," but this should not deter you from asking questions which are vital to your health and the health of your school and community.

INVITING SPEAKERS

Before you decide to take a tour, you might want to invite a speaker from a local grass roots anti--nuclear group or a nearby college or university. Some grass roots organizations have speakers' bureaus while others have one or more individuals who are willing to speak to concerned citizens about the dangers of nuclear power. One of the most effective anti--nuclear sessions I've attended was held in the local high school shortly after Three Mile Island. Some students had invited a professor from Cornell's biology department to spend the afternoon talking about nuclear power. The speaker drew a diagram of a reactor on a black board and proceeded to explain how the fission process works and what might happen if everything does not work right in a nuclear reactor. (There are 50,000 parts in a nuclear reactor.) He explained what had happened at Three Mile Island and what the consequences of a meltdown might have been had the situation within the reactor core continued for just *thirty more minutes*. When he concluded his talk, the room was absolutely silent and when he asked for questions more than one hundred hands were raised.

If you live near a university or college, chances are good that an instructor or professor would be willing to address your students on the dangers of nuclear power. Many activists devote every spare moment they have to talking to groups of concerned citizens about the dangers of nuclear power, and would be more than happy to talk with a group of students.

SHOWING THE SPECK

One of the devices the nuclear industry uses to convince people that nuclear power is clean or less polluting than other industries is for a utility spokesperson to hold up a small pellet (about the size of a rabbit food pellet) and declare that "this pellet is the amount of waste a nuclear reactor will produce within one year's time for each person living near the reactor." When compared to other industrial wastes this appears to be a very convincing argument for nuclear power.

But the industry does not tell people that if this were a pellet of *plutonium* it would be enough to kill *millions of human beings.*

An effective way for a teacher to demonstrate the dangers of plutonium is to place a tiny speck of some material on a microscope and have the students examine the speck. The teacher can also bring a bag to the classroom containing one pound of salt or sugar. After the students have examined the speck, the teacher can explain that, had it been plutonium, what they saw on the microscope would be more than enough to give them lung cancer if it had been ingested into their lungs. And if the one pound bag contained plutonium it would be more than enough, scattered into the biosphere in microscopic pieces, to kill *4 billion human beings.* You might add that a person can carry enough plutonium in a woman's handbag to make an atomic bomb, and that quantities of plutonium are already missing from reactors and have *not been traced.*

If the nuclear industry builds all the plants it has on the drawing boards, 1700 tons of plutonium will be produced by the year 2000. The half life of plutonium is 24,000 years which means that this 1700 tons of plutonium would have to be stored safely for 250,000 years. One *millionth of a gram* of plutonium has produced cancer in laboratory animals.

TALKING 1984

The continuing expansion of the nuclear power industry poses a threat to civil liberties for several reasons. First, nuclear power plants produce plutonium which is the essential ingredient in making atomic bombs.* Opponents and proponents of nuclear power agree that the production of plutonium poses unique problems both for the country in which it is produced and for the entire world. They agree that a quantity of plutonium no larger than a softball is enough to make an atomic bomb that could level a large city, but they disagree on what should be done to prevent terrorists from stealing or buying plutonium. Opponents of nuclear power argue that the best way to protect society from this man-made carcinogen is simply to stop making it. But proponents of nuclear power argue that we must develop specially trained and equipped police units (plutonium swat teams) which will protect nuclear installations and radioactive waste convoys from sabotage or theft. In order to meet emergency situations these men and women would be given a mandate to act which would *supercede* traditional police powers. The Constitution would not apply to the plutonium police. They would be free to act in any way they deemed necessary to see that a quantity of plutonium did not "get into the hands of the wrong people."

Students of history know that a society which delegates such extraordinary powers to a small elitist paramilitary force

A single nuclear reactor produces enough plutonium to make two or three atomic bombs every year. Theodore Taylor, a nuclear safeguards expert from Princeton University, estimates that by the year 2000 enough fissile material will be in circulation to manufacture 250,000 bombs.

is running the risk of giving up certain fundamental liberties. The question is how much freedom or how many of our civil liberties are we willing to relinquish for the high priced electricity generated by nuclear power?

Another reason why nuclear power poses a threat to civil liberties is that enormous sums of money, 17 billion dollars in the United States, have been spent throughout the world to develop nuclear power. Government officials and atomic scientists have staked their reputations on the success or failure of nuclear power while private investors, convinced that nuclear power meant lucrative profits, have continued to back this industry in spite of its dismal safety record and failure to produce "electricity so cheap it could hardly be metered." Government bureaucracies and private industry do not like to admit they are wrong, particularly about something which has taken thirty years and countless billions of dollars to develop. In his recent book, *The New Tyranny,* Robert Jungk describes what happens to those who decide to leave the nuclear club. In Germany, scientists who have criticized the government's nuclear program have been literally frozen out of the job market and relegated to a Kafka—like nonperson status. One scientist has had attempts made on his life. In France, squadrons of police have been sent to guard the reprocessing plant at La Hague from angry citizens who had been told that the plant would manufacture refrigerators or television sets. "The workers at La Hague gave me a glimpse of the world's most frightening working conditions. They have sacrificed not only their health, but their rights to free speech and self—determination. They refer to themselves as 'radiation fodder'—the cannon fodder of the new technology."

In the United States at least two anti—nuclear activists, Karen Silkwood and Michael Eakin, have died violent and mysterious deaths. And, according to the June, 1978 issue of *Critical Mass Journal* 24 utilities "spent more than 2.9 million dollars for outside 'security consultants' above and beyond their on-site plant police." According to CMJ a surveillance operation in Georgia maintained files on private citizens whom Georgia Power considered "subversive," and Virginia Electric Power Company (VEPCO) "asked assembly for official police powers. The state turned down the utility's request." When private utilities begin spying on their rate-payers and lobbying for powers which supercede the Constitution, our civil liberties are in danger.

I would recommend three books to social studies or world history teachers interested in studying the effects nuclear power is having on our civil liberties: *Nuclear Power And Civil Liberties,* researched and written by Donna Warnock and published by Citizen's Energy Project; Robert Jungk's *The New Tyranny;* and *Nuclear Theft: Risks and Safeguards* by Mason Willrich and Theodore Taylor.

I have found that students, particularly teenage students, often feel oppressed and enjoy talking about freedom, civil liberties and human rights. For students who are not interested in the more technical aspects of nuclear power, a discussion of civil liberties can be a good way to demonstrate

the dangers involved in continuing to expand the nuclear in—
dustry.

SOLAR OPTIONS

Convincing students that nuclear power is dangerous is
really only the first step in demonstrating that large, highly
centralized power plants are not necessary to America's economic and social welfare. We must show that people are
capable of building and installing solar green houses, wind—
mills, water—driven generators and homes and factories that
are energy efficient. For example one science teacher in
Western New York State showed his students how to build
"heat grabbers," an idea from *Mother Earth News*. Using
foam rubber and tin foil, the class made several small devices
which were set in windows and produced an air current sending warm air into the room. Each heat grabber cost around
$40 and the students felt a sense of pride when the project
was finished. Another teacher has purchased surplus heli—
copter blades and taken his students on field trips to his
farm where they could help build a windmill. On Cape Cod,
Massachusetts, 22 people calling themselves the *New Alchemy Institute* are studying how we can have "a modern,
advanced, sophisticated, but biologically rather than tech—
nologically dominated society…"*

Teachers interested in showing students that there are
alternatives to nuclear power and that we need not look to
government or private industry for answers to our energy
problems will find *Mother Earth News* and *New Alchemy
News Letter* very helpful.

There are really an unlimited number of ways in which a
teacher can make his/her students aware of the dangers of
nuclear power. The books, films and ideas included in this
section should aide teachers who wish to become teacher—

activists in the struggle against nuclear power. Below is a list
of journals and magazines which are valuable resources in
the struggle for a nuclear-free future. I would encourage
teachers to subscribe to one or more of these.

Critical Mass Journal. PO Box 1538, Washington, D.C. 20013.
Monthly on safe & efficient energy use, and citizen action
against nuclear power.

Friends of the Earth. 124 Spear St., San Francisco, CA
94105. Biweekly magazine published by FOE, *Not Man
Apart,* is devoted to environmental issues, with a "nuclear
blowdown" section describing events, strategies, setbacks &
progress in the anti—nuclear movement.

Groundswell. 1536 16th St. NW, Washington, D.C. 20036.
A resource journal for energy activists and information on
how to oppose nuclear power. Available from Nuclear
Information and Resource Service.

Rain: A Journal of Appropriate Technology. 2270 NW
Irving, Portland, OR. 97210. Monthly.

People and Energy. 1413 K St., NW, 8th. floor, Washington,
D.C. 20005. Excellent bimonthly news magazine. Available
from Citizens' Energy Directory.

New Roots. Box 548 Greenfield, Ma 01302. Science,
Health, Politics, Energy Issues. For the Northeast. Bimonthly.

New Age. 32 Station St., Brookline Village, MA 02146.
Health issues, spiritual and educational concerns, energy
issues and nuclear power.

The Waste Paper. 3164 Main St., Buffalo, N.Y. Concentrates
on the problems of radioactive waste. Good update on
nuclear issues. Sierra Club Atlantic Chapter Radioactive
Waste Campaign. Monthly.

NEW ROOTS, December, 1979.

HOW TO CHALLENGE YOUR LOCAL UTILITY

The more one reads about utility companies, the more apparent it becomes that the American people have been swindled for years, and this swindle has been accomplished in spite of government agencies that were established to protect the rate payers from unfair practises.

In the past five years, the seven investor-owned utilities in New York State have requested 136 rate increases totaling $1.7 billion. The elderly on fixed incomes, the working poor, and the recipients of public assistance have been forced to choose between lighting and heating their homes or feeding their families. But citizen action groups are growing, people are fighting back in a variety of ways. In California an entire town turned its power off to protest high utility rates. In New York State, the town of Massena has fought a long hard battle to form a municipal utility. Through rate strikes, boycotts, the formation of municipal utilities and acts of individual defiance, Americans are challenging the arrogance of privately owned utilities. In small towns and cities alliances are forming, and the long dormant anger is becoming a positive tool for fighting manipulation and exploitation.

Meanwhile, the utilities continue building atomic lemons which fail to deliver the cheap electricity their promoters have promised. It is obvious that the utility companies are in a race against time, trying to outpace the consciousness of the American people. To win this race they are willing to risk our lives.

The material in this section offers both ideas and hope to citizens who wish to challenge their local utility.

THE SHEEP AND I: SOME THOUGHTS ON ALTERNATIVES TO CIVIL DISOBEDIENCE

By John Gofman

> *"The Sheep and I: Some thoughts on Alternatives to 'Civil Disobedience' "*
> *is from John Gofman's IRREVY. In this section Dr. Gofman suggests that*
> *we deduct compensation for the suffering caused by utility companies directly*
> *from our utility bills. Dr. Gofman also calls for wearing black arm bands in*
> *mourning for those killed by nuclear power, and citizen lawsuits against*
> *utilities and their directors for the "grievous injury of mental anguish."*

I am frequently asked, "But what can we *do* to stop nuclear power?" Not all opponents of nuclear power believe in jumping fences, occupations, or blockades at nuclear facilities. Two ideas for other actions come to mind: both could provide mechanisms for educating more people about the deadliness of nuclear power, for raising awareness about the inalienable right to life, and for manifesting a fundamental principle of morality: *Individuals must be held responsible for what they do and help to do . . .* the principle briefly advocated by this country at the Nuremberg Trials.

MENTAL ANGUISH

There simply can be no doubt that one of the most serious forms of injury to humans is the induction of mental anguish. Anguish is a legally recognized form of personal injury. It comes up regularly in divorce cases as "mental cruelty."

Humans in general are horrified to think that their action may have caused the death or injury of other humans. Without a doubt, this concern is so well developed in many humans that they will suffer a severe state of mental anguish from knowing that their actions have helped to kill other humans.

Recently, the electric utility industry and its officials have taken willful action which induces mental anguish of the most severe form in a large number of utility customers. Millions of persons who need electricity in their homes and places of business are forced, by the monopoly arrangement of electric supply, to purchase their electricity from companies which have gone into the human murder-business by constructing and operating nuclear power plants, for which customers must pay.

That nuclear power plants *are* committing premeditated random murder is undeniable. Even the Environmental Protection Agency acknowledged in 1975 that nuclear power will kill hundreds of members of the public every year, even if everything goes perfectly!* While the EPA has grossly underestimated the number of victims, the Agency has effectively confirmed that premeditated random murder will occur.

In 1978, a major advance in government honesty was made when the Nuclear Regulatory Commission finally admitted there is *no* known safe dose of ionizing radiation, no "threshold."*

Thus, people who need electricity have been forced against their own desires and convictions to participate in the process of murdering innocent humans, toward whom they bear no malice. And this has induced a severe state of mental anguish in countless thousands—perhaps millions— of such persons. Indeed, in the hope of relieving the mental anguish, many have even sacrificed high earning capability in order to be active in the anti-nuclear movement.

Possible Remedies

There are at least two possible remedies to compensate for the grievous injury of mental anguish:
a.) A swarm of citizen *lawsuits* against utilities and their directors for mental anguish and punitive damages;**
b.) A mass *direct-action* campaign against utilities to obtain at least partial compensation for the anguish.

Suppose the victims suffering from mental anguish were to get together and decide that an appropriate compensation

---

* U.S. Environmental Protection Agency. *Draft Environmental Statement for a Proposed Rule-Making Action Concerning Environmental Radiation Protection Requirements for Normal Operations of Activities in the Uranium Fuel Cycle.* Office of Radiation Programs, EPA, Washington D.C. 20460, May 1975.

---

*Confirmatory documentation:
U.S. Nuclear Regulatory Commission. *July 31, 1978, SECY-78-415, Policy Session Item from Robert B. Minogue, Director, Office of Standards Development, to the Commissioners.* Subject: Further Actions to Control Risks Associated with Occupational Radiation Exposures in NRC-licensed Activites. See especially page 11 of its "Enclosure B," in which the Office of Standards Development urges discontinuation of the term "permissible dose" because it has been *mis*interpreted to mean "safe," when in fact, "Considerations of the linear hypothesis indicate that some risk is associated with any dose of radiation, however small."

It is a pleasure to give credit where credit is due. In a fine letter to Dr. Gofman, September 11, 1978, Drs. Minogue and Karl Goller (of the same office) conclude that, ". . . tough issues have to be faced in dealing with non-threshold pollutants; we think there is growing awareness that radiation is only one of these."

**It may be pertinent that on August 11, 1978, the famous attorney, Melvin Belli, filed a lawsuit against the Bank of America on behalf of a client who could not get back $20,000 deposited in the Bank's Saigon branch shortly before the North Vietnamese took over the city. In addition to claiming the $20,000, Belli is suing the Bank for $1,000,000 for "emotional distress" (plus $3,000,000 additional for punitive damages).

for their suffering would be $100 per month. If they live in that "service region" on the average for five years, the amount would be $6,000—a very modest sum for such a grievous injury.

Suppose, then, that these electric customers agree to start sending bills to the utility for their injury. Thus someone receiving a bill for $40 from the utility, instead of paying the bill, simply sends the utility a bill for $60— the difference between what the utility owes him/her, and what he/she owes the utility.

In no way would this action deny the justice of paying for electricity. Rather, this action is simply a recognition that individuals have rights, not just utilities, and that there is a *two-way* transaction going on. The utility is selling power, and I am buying it, but the supplier (there is no choice of supplier) is involving me also in the murder-business, which changes the terms of the transaction.

Now it could happen that some utilities might be so unreasonable as to deny they owe *any* mental anguish payments, and they might even act rashly to threaten disconnecting delivery of electric power to the victims, even though this would be manifestly unfair.

In some states (e.g., New Hampshire), power customers have a right to a conference with the utility at the utility's office, if they think termination of service would be unjustified. And grossly manhandled customers who are not satisfied with the results of that conference may request a conference about the matter with a staff member of the Public Utilities Commission prior to termination of service. (Details from the Granite State Alliance, 83 Hanover St., Manchester, NH, 03101)

When and if utility companies appear deaf in hearing the just claims of their customers, involvement by an increased number of customers could materially assist the utility in overcoming its hearing defect.

VISIBLE MOURNING

It is, of course, difficult to suggest simple and adequate relief of the great mental anguish being suffered by people who are forced against their wills to participate in the process of randomly murdering other humans in this and future generations. One small, but perhaps important, measure which could help provide relief would be a *public expression of the mourning* felt for those being murdered by nuclear power.

Since the mental anguish is a very deep ethical matter, it is essential that the mourning be at a level of dignity commensurate with the deep seriousness of the problem. It should be quiet, dignified, but unmistakable to the community. Historically, a highly dignified method of expressing grief for the loss of another human has been the wearing of a black armband by those in mourning.

A simple, black armband would do very well. However, it would be helpful for the community to know the source of the sadness, so it would be appropriate for the armband to identify the utility company engaged in nuclear murder in each specific locality where the armbands are being worn. Surely it is important to convey to those conducting the random murder, the feelings of those suffering the mental anguish therefrom. Vigils near the homes of utility directors and officers would increase their awareness.

If we are asked why we mourn nuclear power victims without also mourning fossil-fuel victims, we would have the opportunity to educate people *also* about the victims of fossil-fuels, plastics, lead-poisoning, Kepone, asbestos...

Some of those older killers are intricately entangled now in our way of life, whereas nuclear power is not. In fact, all the nukes in the U.S., combined, still contribute just 3½% of our total energy supply.

The fact that some older technologies are turning people into basket-cases and corpses, is one of the strongest arguments for the good sense of *preventing* another tragedy from radioactive poisons. Visible mourning might help, as much as or more than fence-jumping.

POWER TO THE PEOPLE:
TOOLS FOR UTILITY ORGANIZERS

By Claudia Comins

"Power to the People, Tools for Utility Organizers," is from the April, 1978 issue of POWER LINE, *and lists organizing resources that can be valuable to utility activists.* POWER LINE *is a monthly publication devoted to utility issues with useful information about what people are doing to challenge their local utilities.* POWER LINE *is available from: Environmental Action Foundation, 724 Dupont Circle Bldg., Washington, D.C. 20036. Subscription rates: $6.00 for one year.*

Citizen action groups are growing almost as fast as utility bills are rising. This spring, utility organizing is well underway in every state in the country. Environmental and consumer groups alike are rallying to oppose rate increases and to stop construction of nuclear power plants and high-voltage power lines.

As those groups grow larger and more diverse, they often find themselves in need of advice on different organizing methods. A group which came together to fight a rate increase may well decide to stay together to work on reforming rate structures or promoting conservation—issues demanding special resources and expertise. Simultaneously, members may start to feel financial pressures and decide to formulate a long-term plan to cover upcoming costs.

Whatever the size and shape of your organization, you may soon be looking for advice on new strategies and funding sources. Fortunately, a vast array of up-to-date books, methodologies and even training programs is currently available to groups working on utility issues.

Fund-raising is one of the most crucial aspects of successful utility organizing. With sufficient funding at its disposal, an organization can hire a much-needed staff person, print flyers or hold a press conference when a situation demands immediate action.

Some publications give particularly useful advice on how to develop a stable fund-raising plan. The Youth Project's *Grassroots Fund-Raising Book* is an excellent guide for those membership groups exploring a wide spectrum of options. It is packed with ideas, how to's and detailed bibliographies on specific subjects. Author Joan Flanagan discusses such activities as benefits, auctions, ad books, marathons, speakers' bureaus, membership, canvassing, publicity and funding sources.

To supplement the book, Flanagan and others around the country will soon be conducting a series of grassroots fund-raising workshops. Each is designed specifically to assist membership organizations in gaining financial self-sufficiency.

Fund-Raising in the Public Interest, prepared by Ralph Nader's Public Citizen, Inc., is another useful handbook for those looking into funding sources. It focuses on three specific fund-raising methods: direct mail, professional canvassing and marathons. Included are detailed descriptions and a bibliography for further research.

More and more large-scale organizations—most notably Massachusetts Fair Share, Illinois Public Action Council and California's Citizen Action League—are depending on door-to-door canvassing for their funding needs. Local groups may also want to look into some form of canvassing, geared to their size and needs. Canvassing can be a risky, problematic business, however, and groups first should thoroughly examine its benefits and drawbacks.

The possibility of receiving grants from church groups must not be overlooked. The United Church of Christ, the Methodist Church and the Catholic Church's Campaign for Human Development are well-known funding sources for utility and other community groups. Local and regional church offices in particular should be approached to support community organizing efforts. Foundation or government agency grants are also possibilities. For a useful bibliography of such sources, see *The Grassroots Fund-Raising Book.*

Another, less obvious, fund-raising technique is that of coalition building. Because of the expanded membership they offer, coalitions aid in efforts like benefits, raffles and speakers' bureaus. Members can help by donating free office space, telephones, printing and staff time. In addition, information can be passed along through union newspapers, church bulletins and senior-citizen, minority and environmental newsletters.

As utility groups prepare organizing campaigns, they may want to discuss tactics with others who have been involved in similar experiences. Training schools have been set up a across the country to provide just that kind of dialogue among organizers. Some schools even have travelling workshops and will offer on-the-job training with field placements. Others can act as consultants on an individual basis.

Some training centers and consultants are also experienced in group facilitation, inter-organizational structure and maintenance. This more personal side of organizing can prove especially useful for growing groups.

Here at Environmental Action Foundation, the Utility Project's Clearinghouse tries to answer the technical needs of utility activists and **organizers**. We offer three distinct kinds of information:

State Utility Activists List: We continue to enlarge and update a list of individuals and organizations active in utility issues.

National Utility Experts List: This list, broken down into subject categories, includes economists, engineers, lawyers and researchers with technical knowledge of utility issues. Groups use this list to find expert witnesses for rate hearings or for general information and advice.

Utility Action Guide: Here we take note of current reports, books, fact sheets and information packets useful to those wishing to pursue specific utility issues and needing some start-up information.

The Utility Clearinghouse itself can refer inquiries to the appropriate government agencies, public-interest and research groups or to experts on given subjects. Our chief aim, however, is to stay in touch with utility issues concerning local groups and to keep the utility network growing.

The Utility Project is currently preparing an information packet on utility organizing which will include detailed information and resources. Please contact Claudia Comins here at EAF if you can provide examples or suggestions on such topics as tactics, training and fund-raising.

REFUSING TO PAY

By Susan Blake

"Refusing to Pay" is from People's Energy Primer, *a special issue of the Syracuse Peace Council's Newsletter. In this article Susan Blake, Activities Coordinator of Peace-Smith House, a peace and social justice community group in Massapequa, New York, discusses some of the pros and cons of rate withholding campaigns.*

Rate withholding is a defiant step which one can take from home in concert with others to protest rate hikes, nuclear power, inequitable rate structures, and the lack of truly democratic channels for participation in decisions affecting our everyday lives. The organization(s) sponsoring a campaign to withhold must minimize risk of service loss or legal recrimination through careful research and strong, constant coordination and communication. Even with broad-based support, the organizing is sure to be long, hard and frustrating.

The rate withholding tactic is probably most effective when done in combination with other utility actions. Just as most people who want to make a stand against nuclear power are more likely to attend a legal rally than to do an act of civil disobedience, most outraged consumers are more likely to sign a petition calling for no rate increases than to hold that increase back from their electric bills. A local campaign to oppose a rate increase or a particular part of the rate structure should offer a range of options on which people can act so as to involve the broadest possible spectrum. Door-to-door canvassing could request an individual to sign a petition, to write a personal letter to the Public Service Commission (PSC), to attend a public meeting, to speak at a PSC hearing, and then, if the person seems quite dedicated, to hold back (partial) payment from her/his electric bill. Few people would be put off by an appeal to sign a petition demanding that their electric bill not go up, but some might not listen if called upon to do the more radical act of withholding.

In my judgement, the Long Island Safe Energy Coalition's (LISEC) first year of rate withholding organizing has suffered from lack of planned coordination. Often other anti-rate hike actions were being done concurrently by many of the very activists working hardest on the rate strike effort.

At the time of this writing, LISEC has about 275 families pledged to withhold any rate increase as soon as 999 others have pledged to do the same. (LILCO is currently asking for an 18.5% rate increase, most of which is earmarked for financing construction of the Shoreham nuclear plant.) We stopped pressing the withholding pledge campaign after the first few months of organizing when the majority of the pledges were collected. During that time, we got excellent publicity and made new contacts with diverse kinds of consumer groups, many of whose members were sympathetic and anxious to help in other ways but did not want to withhold. On a series of consecutive Saturdays in winter '77 when we had people across the island scheduled to canvass with the pledge appeal, severe ice storms thwarted our efforts. (LILCO felt the wrath of these storms worse than we did with massive electrical outages due to fallen lines!) So for a variety of reasons, we rate strike organizers began to put more energy into other utility actions. But pledges continue to trickle in, and these committed electric consumers can be called together by phone, mail and meetings to act on utility issues, including perhaps eventually, when the time is right, to withhold.

It's left to us now to follow up on the work we've started. I think it's been worth the effort.

So, if you should decide to try it, see the following page for a few of the organizational hurdles you are likely to face.

THE RESISTER'S GUIDE TO WITHHOLDING UTILITY PAYMENTS	
What to withhold?	A rate increase (can be used to raise all rate structure issues as well as to dramatize direct economic hardship). Percentage of bill going towards nuclear plant construction and/or operation (or towards high voltage lines or phantom taxes or ...). The fuel adjustment factor. A symbolic, standard amount for everyone.
When?	After a goal amount of pledges have been collected. At a given date, with or without collecting pledges.
For how long?	One payment. For a sustained period.
What to do with the money? (Organizers must warn potential withholders that they will, in all likelihood, eventually have to pay).	Individual withholders save it for eventual payment. Put it into escrow, or into a fund for organizing expenses, or for alternate energy and conservation, or for utility payment relief. (Placing the withheld money elsewhere de-emphasizes the reality that most people simply don't have what's being asked of them by the utilities.)
How to respond to threat of shut-off? (These options must be clearly outlined for potential withholders. Strength in numbers and counselling in optimal use of PSC law do not guarantee protection from possible consequences of withholding.)	Pay. (Avoids any risk of legal shut-off.) Appeal your complaint to the PSC, and then pay. PSC law mandates advance notice, including a personal conversation, before shut-off. During this time, a consumer may request a personal investigation of her/his complaint. Shut-off is illegal during this process, although pressure may be exerted for partial or complete payment. Correspondence with the PSC on intent to withhold should be done individually, not by form letter or by the sponsoring organization(s). Refuse to pay and use the occasion to draw publicity (e.g. demonstration of support at house where shut-off is imminent, nonviolent blocking of shut-off, use of alternative sources).

CALIFORNIA TOWN PULLS THE PLUG

By Deborah Schoch

"California Town Pulls the Plug" is the story of one small town's resistance to ever rising utility rates. Outraged by bills that doubled and then tripled within one year, the entire town of Westwood "pulled the plug" in a week long boycott that resulted in an investigation of the utility's voltage levels and charges.

A retired laundry operator in Westwood, Calif., with a monthly income of only $300 in Social Security and veterans' benefits, was mailed a $113 electric bill one month this winter.

Another Westwood resident, unemployed with a family of four, found himself paying a bill for $80—one-third of his total income.

Finally Westwood's population lost all patience with its local electric company. On Feb. 12, 1978, this town of 2,500 located in California's High Sierras, began a voluntary weeklong power blackout to protest soaring electric bills. Virtually everyone went without lights, hot water, stoves and television sets. At night, candles and kerosene lamps flickered in windows along the downtown streets. The only bright lights in sight were those illuminating the local headquarters of California-Pacific Utility Co. (CPU), Westwood's wayward utility.

Word of the boycott spread fast. Television crews rolled into town, and photos of residents crowded around wood stoves made newspapers across the country. Even before the blackout's end, Westwood had become the national symbol of consumers standing up to their electric company.

"We figure it's a patriotic move—like the Boston Tea Party," explained Westwood native Pauline Asmus. She pointed out that the boycott was scheduled to run from Lincoln's birthday to Washington's. "Lincoln freed the slaves," she added, "and the way rates are now, we're slaves to CPU."

Judging from the looks of CPU's electric bills, Westwood has good reason to be angry. Those bills have doubled or even tripled in the past year alone. The town's economic backbone is the seasonal lumber industry, which shuts down each winter, and townspeople can hardly afford electric bills ranging from $50 and $100 a month to more than $200.

CPU is not known for accurate billing procedures. Bills are often sent to the wrong names or addresses; recently a one-day, $25.52 bill for a private contractor was switched with the $4877.15 monthly bill for Westwood Elementary School. And at least 116 of Westwood's 850 families got January bills with overcharges of up to 55.3 percent.

"We sent protest petitions to the utility and never heard anything," said one exasperated billpayer. "They treated us as if we had no voice at all."

CPU countered with claims that townspeople didn't understand how their bills were figured. "If they're paying more, they're using more," maintained one utility official.

Then the boycott prompted an investigation of the utility's voltage levels. Checks by both the state Public Utility Commission and *The San Francisco Chronicle* disclosed that CPU was supplying Westwood with a lot more power than it needed. Voltage levels in 110-volt lines ranged from 124 to 127 volts—an excess capable of causing up to a 10 percent increase in the amount of kilowatt-hours consumed. CPU confirmed the findings, and a state investigation of this and other problems with Westwood's service is expected shortly.

Whatever the end result of Westwood's determined stand, it has already built a coalition of short-tempered electric customers everywhere. Calls of sympathy were received from every part of the U.S. as well as Canada, Mexico, England and Australia. And dozens of callers from towns as diverse as St. Louis and San Jose vowed that they, too, were pulling the plug in a gesture of support.

SHORT-CIRCUITING HIGHER RATES

By Fran Weisberg

"Short-Circuiting Higher Rates" is from People's Energy Primer, *a special issue of the Syracuse Peace Council's Newsletter. It describes some of the reasons for switching from private to public power and offers a few suggestions for working toward public power in your community.*

The Syracuse Peace Council's Newsletter is an excellent resource for energy and peace activists. Subscriptions are $6.00 per year and can be obtained by writing: Syracuse Peace Council, 924 Burnet Avenue, Syracuse, New York 13203.

Residents in Rochester, N.Y., pay about $23 every month for 500 kilowatt hours of electricity. A few miles away in Fairport, N.Y., residents pay only $9.32 for the same amount of electricity.

In Rochester, utility customers are facing a 20% electric rate increase request. In neighboring Fairport, rates have changed only once in the past ten years, and that was a rate decrease.

How can this be?

Fairport has a municipally-owned utility, operated without profit for the benefit of its citizens.

The goal of Rochester Gas & Electric Corporation (an investor-owned utility) is profit—not energy at the lowest possible price.

A public electric or gas system—there are 48 in New York State—won't solve energy shortages or end debates over the best possible energy sources. But it's a beginning, and the only way to insure an energy path for the future based on need instead of profit.

Public power requires a strong community movement, a government-financed study, and voter approval.

A public power system is a gas or electric system owned and operated by a city, town, county, state, or special district. The municipality or district can own the generating system, or just the distribution system, buying its power from the cheapest wholesaler.

A non-profit public power system can provide electricity at rates 30 to 50 percent cheaper than investor-owned utilities, according to the U.S. Federal Energy Regulatory Commission (FERC).

More than 14 percent of all Americans—30 million people —power their homes through 2,224 public power systems across the country. Systems vary in size from those owned by small towns to statewide systems such as Nebraska's state-owned electric utility.

The three major reasons public power systems save customers money are:

1. A non-profit utility has no stockholders and pays no dividends.

2. A public utility can borrow money at lower interest rates because the bonds it issues are tax-exempt.

3. Public power systems are operated and organized more efficiently, according to a FERC report, and costs are lower per kilowatt sold.

The first step in any public takeover of a privately-owned utility is a feasibility study, authorized by the local governing body and done by an independent, professional engineering firm. This study examines the legal, financial and engineering aspects of a proposed public system to ascertain whether a community would save money by owning its utility.

The cost of the feasibility study varies. Buffalo's City Council authorized $100,000 to study the feasibility of taking over both its gas and electric systems. Massena, N.Y., a town of 14,000 on the Canadian border, spent $30,000 for a study that was the first step in acquiring its electric system from Niagara Mohawk Power Company.

If the feasibility study shows a public system would be less expensive, then the community decides by ballot if the investor-owned system should be bought.

A strong, well-organized mass movement to educate the public is imperative to counter the private utility's campaign to maintain control of the service area. The utility's campaign against public takeover shouldn't be underestimated. RG&E spent more than $20,000 in Hilton, N.Y., a town of less than 2,000, to thwart a public power referendum there, and based its campaign on scare tactics about unemployment and power blackouts that would occur if Hilton owned its electric system.

If the proposal is approved by voters, then the municipality begins condemnation proceedings on the utility's property within municipal limits. The court usually must decide what is a fair price for the property, and the city or town issues bonds to purchase it.

Now that you own your own electric system, where do you get your power?

Energy sources and prices are outlined in the feasibility study and include shopping for the cheapest power from private utilities, or obtaining low-cost hydro power from the Power Authority of the State of New York (PASNY).

PASNY offers cheap power from its Niagara Falls Hydroelectric Power Project. By federal law, it is mandated to make 50% of that power available to public power systems. Currently, only 12% of that electricity powers all of New York's 48 public systems. But the state's public systems have initiated a court suit to wrench more PASNY power from the private companies.

It's important to remember that cheap PASNY power isn't the only way a public utility saves money. Public systems save money by removing profit, stockholders' dividends and high administrative and advertising costs from their operating budgets.

With electric rates skyrocketing every year, feasibility studies on public power have been commissioned from Buffalo to Westchester County, in New York City, and in Hudson Valley. Many other communities such as Geneva, Rochester and Oswego are all actively involved in public power. In your community, contact the state People's Power Coalition or any affiliate group.

POWER TO THE PEOPLE OF MASSENA!

By Susan Madsen

This article comes from the September, 1979, issue of the Syracuse Peace Council's Newsletter. It gives the background on the struggle to form a municipal utility in Massena, New York. Today there are 2,000 municipal utilities in the United States and the number is increasing.

Six years ago, the upstate town of Massena (pop. 16,000) began waging a valiant battle for economic independence from one of New York State's utility monopolies, Niagara Mohawk. It had been the first time in over 30 years that any New York municipality has attempted to break away.

Massena's goal has been to buy power directly from the Power Authority of the State of New York (PASNY). PASNY power is cheaper because PASNY is a public corporation (no large dividends to pay stockholders as in private, investor-owned utilities like Niagara Mohawk).*

It appears that the last straw for Massena was Niagara Mohawk's attempt to raise electric rates by 23.5% in 1974. Although the company was granted only a 7.5% increase, during the early municipal campaign, Niagara Mohawk then turned around and asked for yet another 16%. John G. Haehl, President and Chief Executive Officer of Niagara Mohawk, announced at a stockholders meeting a few months before the vote that, "regular rate increases in the years ahead are now a fact of life." This surely added fuel to the fire in the town's determination not to be victims of Niagara Mohawk's profiteering. In the previous five years, before the referendum, Niagara Mohawk had raised the rates to its customers in excess of 30%.

A feasibility study was done which proved that municipalization could save the town power costs in excess of $6 million between 1974 and 1984.

Niagara Mohawk, realizing what a victory in Massena could mean, began an avid campaign to prevent the takeover. $290,000 was spent on such tactics as full-page newspaper ads, door to door campaigns, pressure on Niagara Mohawk employees to influence neighbors and relatives, an open house to display new equipment. None of the equipment, however, was ever used in the town of Massena. And, on voting day of the referendum, Niagara Mohawk provided a

fleet of cars which transported people back and forth to the voting place and allegedly paid poll watchers $50.00 a day. An interesting point to note here is that the rate payers are financing, through their bills, Niagara Mohawk's opposition to an effort which hinders rate payers' pursuit of affordable, equitable energy rates.

During Niagara Mohawk's campaign, they made some of the following statements to sway people against municipal power: "Municipal operations would raise taxes; the community would be helpless in the event of a disabling storm; Massena could not staff a municipal system; engineering studies underestimated takeover costs; the municipal system could not pay union wages and would therefore put people out of work."

The local United Auto Workers didn't believe *any* of these allegations. They went out and raised over $2,000 to wage a campaign to meet Niagara Mohawk head-on. Members collected statistics and facts to disprove Niagara Mohawk's statements. The United Auto Workers assured that the truth reached the people and finally moved the people out to vote on referendum day, May 30, 1974. By a margin of 3 to 2, and with 74% of the voters coming out to vote, the referendum was passed to acquire the facilities of Niagara Mohawk Power Corporation and establish its own municipal system.

Why did Niagara Mohawk spend in excess of $2 million on their campaign to deceive the people of Massena? An article in the *New York Times* which appeared right after the referendum vote may answer this:

> If Massena goes, can Niagara Falls be far behind, and if that happens, will Buffalo, Rochester and Syracuse be far off?

More than being just a precedent for future municipal campaigns, Massena is a very key issue in the self-determination of a town to choose autonomously how it will conduct and regulate its energy use.

And, what does John Haehl have to say? In 1978, in the Annual Stockholder Report, responding to Massena's "bid

---

*PASNY power is also cheaper because PASNY, as a public power system, can borrow money at lower interest rates through tax-exempt bonds. The PASNY power that Massena is seeking is particularly inexpensive because it is from the St. Lawrence hydro station.

to create a municipal electric system in expropriating Niagara Mohawk facilities:"

> We will protect the interests of stockholders to the fullest extent by continuing to exert absolute opposition to this effort—or any similar takeover threat.

Since 1974, counsel for Massena has attempted to negotiate a price for the acquisition of Niagara Mohawk facilities. Because there had been no agreement on the cost, numerous court battles have ensued. However, Massena has won each battle along the way, until recently when Niagara Mohawk refused to transmit the power that was needed by the town of Massena until a federal court case was decided. Even though Massena was given the go-ahead by the State's highest court, Niagara Mohawk had decided to bring its case before an even higher court. Federal Judge Howard Munson granted Niagara Mohawk a temporary injunction blocking Massena from starting.

How do the people of Massena feel? According to the town's attorney, Eugene Nicandri, "The sentiment in the town is that Massena will prevail."

The issue will be won when, in the courts, Niagara Mohawk realizes that not only Massena, but other people defending economic justice will not tolerate the rule of our lives by utility monopolies. We must work together angrily, visibly and in large numbers to make affordable energy and peoples' control of energy a reality.

PUBLIC POWER BITES THE BIG APPLE

Can New York City's nine million residents form a municipal utility which will end the reign of Consolidated Edison in the Big Apple? According to POWER LINE work toward this goal has already begun.

"If there was a competitive choice, given Consolidated Edison's apparent current standing in the eyes of New York consumers, Consolidated Edison probably wouldn't have any customers," concluded a special New York City commission investigating the blackout that paralyzed the city in July. The commission's 400-page final report, released Dec. 1, 1977, aims a blistering attack at Con Ed's management and service.

A great many New Yorkers agree wholeheartedly with those conclusions. They continue to pay the highest electric rates in the nation for what is probably the nation's worst electric service. Certainly they would welcome an opportunity to put Con Ed out of the electricity business. And if organizers now at work in New York are successful, city residents will get that opportunity.

In short, a push for public power is underway in the Big Apple. A coalition known as New York City POWER (People Outraged with Energy Rates) is slowly but surely educating New Yorkers about the advantages of a publicly owned utility. Among the coalition's 25 member groups: the United Church of Christ, the local chapter of the National Council of Churches, the Manhattan Council on Housing and numerous neighborhood and tenant groups.

Buying out Con Ed may at first seem audacious or even ludicrous. Con Ed, after all, is the most complex, problem-ridden utility around—and the present New York City government is hardly in a financial position to purchase a utility.

But POWER has come up with some convincing arguments why New York should thoroughly investigate public power. According to Richard Schrader of the statewide Peoples' Power Coalition, a member group of POWER, the most tangible benefit would be a chance to save jobs by retaining the small and medium-sized industries now fleeing the city.

"We've lost breweries, bakeries—sometimes they go just across the river, where they pay half the cost of electricity," noted Schrader, a Bronx resident. Other potential benefits: lower residential rates and better incentives for holding down costs and exploring decentralized energy sources.

Arguments like these have already generated support for a public power feasibility study from 12 of the 35 members of City Council and the borough presidents of both Manhattan and the Bronx. Under state law, municipalities considering public power must first undertake such a study and then put the question to referendum. If all goes well, POWER estimates, a formal feasibility study—costing approximately $200,000—could be commissioned in the late spring of 1978, with a vote as soon as 1980.

Exactly why has New York City become a prime site for a public power showdown? The most obvious reason is the July blackout, which may have permanently destroyed customers' faith in Con Ed's service ability. A multitude of investigations, including the commission report released in Dec. 1977, blamed the blackout on the utility's incompetence.

Also in New York's favor is its access to a statewide public power source, the Power Authority of the State of New York (PASNY). According to POWER, PASNY could purchase the Con Ed system, which could then be administered either by PASNY itself or by the city's Planning Board. Con Ed executives have mentioned a $5.5 billion price tag—but public power advocates are estimating a purchase price of half that amount.

Still, as Schrader puts it, the city's public power organizers face a "massive, massive campaign." At worst, their efforts may heighten awareness of utility issues and the virtues of public power. And at best? Residents of the nation's largest city will own and control their own power system.

CITIZENS LAUNCH NATIONAL SHUTOFF CAMPAIGN

By Claudia Comins

In Louisville, Kentucky a group of senior citizens has started a campaign to see to it that no one else dies from having their electricity shut off in the middle of winter. This is a very important step toward ending the domination of the utility companies.

December began with its customary grisly toll in 1978 as nine members of a low-income family in Houston burned to death after the candles they were using for light ignited natural gas leaking from a line they had apparently re-connected illegally for heating purposes. Blowtorch-like flames then gutted their house.

The details were unique, but in a sense the case was depressingly familiar. In the last several winters, the National Council of Senior Citizens contends, at least 200 people have died after their gas or electricity was shut off, and thousands have suffered varying degrees of privation.

This year (1979), something may be done about it. The newly formed Citizen/Labor Energy Coalition has launched a several-pronged campaign on the state and national levels to demand that regulatory commissions enforce a "no shut-offs" policy this winter. Coalition affiliates in 20 states have held press conferences and rallies on the issue; members have presented petitions to city councils; and C/LEC affiliates are demanding that their state regulatory commissions take the lead from Wisconsin's commission and ban shut-offs in cases where health may be affected.

The Citizen/Labor Energy Coalition was formed earlier this year to counter the enormous power of the energy companies and to insure that energy continues to be available and affordable for consumers. Member groups include citizen action groups, unions, churches, minority organizations, environmental groups and senior citizen organizations.

The campaign first got rolling in Louisville, Ky., where senior citizens, church people and labor groups marched on Louisville Gas & Electric to demand a winter shut-off ban. Rallies followed involving the Neighborhood Energy Alliance in Minneapolis, and Utility Consumers United in Milwaukee. The Champaign, Ill., city council was persuaded to declare 11/21/79 (kick-off day for the campaign) "No Shut-Offs Day."

On the national level, the Coalition has found an ally in Sen. Ted Kennedy (D-Mass.) who promises to present information about suffering related to shut-offs—and instances of abuse suffered by consumers —to Congress.

Moreover, the Coalition has grasped the opportunity to use the new Federal Energy Act. Organizers pressured the Department of Energy to write to the states recommending the implementation of new federal standards concerning shut-offs. DOE Undersecretary David Bardin agreed, penning a letter on 11/15/79 in which he expressed hopes that commissions will either complete or undertake an "effort to adopt rules precluding termination of service during the coming heating season solely on the basis of inability to pay."

Working with C/LEC is the National Consumer Law Center, which will answer legal questions regarding the shut-off moratorium from its office at 11 Beacon Street, Boston, Mass. 02108 (617) 523-8010. The Coalition has also compiled up-to-date information on shut-offs, sample PUC petitions, state surveys, and suggested strategies and tactics. For more information, contact the regional representatives in your area.

CON ED MUST BUY CITIZEN WINDPOWER

By developing and building alternative energy projects. Americans are taking control over their own lives in small but very important ways. In New York City a group of lower east side residents built a windmill on the roof of their building and "won a smashing victory over Consolidated Edison" when the utility was forced to buy power generated by the citizens' windmill.

Residents of an apartment cooperative on New York's Lower East Side won a smashing victory over Consolidated Edison Company in May 1977. The state's Public Service Commission (PSC) ordered the utility to buy power generated by the citizens' rooftop windmill. Con Ed had tried to disconnect service to the building because the wind generator, which was hooked into the company's power lines, caused the customers' electric meter to run backwards.

The PSC's decision sets an important precedent for promoters of decentralized energy systems who frequently encounter resistance from the utility monopolies.

When the windmill was installed in the fall of 1976, Con Ed executives threatened to shut off the building's electricity supply if the mill continued to feed power into the company's grid. The officials said that power from the two-kilowatt windmill might cause some damage to its 10-million kilowatt system.

The half-century old, three-blade device supplies lighting for the building's common spaces such as hallways, as well as power to pump the solar water heater. When the windmill's output is insufficient, Con Ed supplies back-up power. And when the windmill produces a surplus, the electricity flows backwards through the meter into Con Ed's system.

The Public Service Commission directed Con Ed to pay 2.3 cents for each kilowatt-hour (kwh) it buys from the windmill, based on the company's fuel cost savings. In contrast, Con Ed charges its customers about 9 cents per kilowatt-hour. That means Con Ed will owe the co-op about 46 cents per month for the anticipated 20 kwh surplus from the windmill. The buildings residents will receive additional savings, however, since each wind-generated kilowatt-hour they use knocks nine cents off their Con Ed bill.

Still, the PSC's ruling is a mixed blessing. Con Ed will be allowed to charge the cooperative a special eighteen dollar monthly fee for monitoring the system, which will wipe out most of the windmill's savings. Nevertheless, the decision opens up utility operations to direct competition from their customers.

Already, several other New York residents have expressed interest in replacing Con Ed's juice with a rooftop wind machine. With the highest rates in the nation, Con Ed has good reason to be worried.

OPPOSING NEW POWER FACILITIES

By Environmental Action Foundation

The following information is from "How To Challenge Your Local Utility" by the Environmental Action Foundation. The information compiled by EAF is invaluable to anyone interested in how the utilities work and to those who want to become active in the anti-nuclear struggle. For more information on challenging your local utility, write: Environmental Action Foundation, 724 Dupont Circle Bldg., Washington, D.C. 20036.

Historically, most power companies have been able to build new plants and transmission lines wherever and whenever they wished. The right of *eminent domain* enables utilities selling electricity, natural gas, water, etc., to condemn land they require for their operations. In recent years, environmental awareness has caused government officials and the public to give the expansion of power facilities increasing scrutiny. In many states, utilities must obtain permission from government agencies to build new power facilities. Environmental protests are being heard more frequently before county commissions, state and federal agencies and in the courts. Power company executives encounter public opposition to practically every new power facility they propose. If the utilities were to begin serious energy conservation efforts and install the best available pollution control equipment on all their new plants, they would have fewer siting problems. Until then, environmentalists must continue to oppose the power industry's blind expansion. Certainly some new power facilities will always be necessary, but the vast majority of new plants and transmission lines deserve critical evaluation by environmentalists.

Opposing a new power plant doesn't necessarily mean getting rid of it. Environmentalists involved in a power plant controversy should consider alternate sites and the need for better pollution control equipment. Citizens should certainly investigate whether or not the plant is needed; even if construction is not halted or delayed, the controversy will educate government officials and the public about the need for energy conservation and environmental protection.

Getting Started

When you first find out about a new power plant, you should generate public opposition as soon as possible. Perhaps you can influence the utility's decision on site location or pollution control equipment before its plans have been finalized. Utilities know that lawsuits can cause monumental headaches and long construction delays. If utility executives anticipate a large public outcry, they may build the plant elsewhere (hopefully in a more appropriate location), or they may decide to include cooling towers or better air pollution controls in the plant's design. The purchase of large blocks of land on a waterway and the appearance of survey crews is usually the first evidence that a company is planning a new power plant. If the company has already publicly announced the plant (including capacity, fuel source, design, etc.), it is probably too late to scare it away. It has already committed too much time and money to the project.

Citizens facing a proposed transmission line should also drum up opposition as early as possible. However, there are a number of considerations which make a power line controversy different from plant siting. The utility need not purchase any land; it needs only an *easement*, giving it the right to cross the property with a power line. It is essential that affected citizens work together since power companies often try to play one property owner against another. Unfortunately, many property owners are more concerned about how much money they can get for the easement than about what happens to the countryside.

There are several possible alternatives for locating a power line. Sometimes there will be a nearby right-of-way for a railroad, highway or existing transmission line. If a line is proposed through a park or over a mountain, you might suggest a route where it would be less noticeable. While prohibitively expensive over long distances, placing power lines underground may be a viable solution in residential areas. (More information on power lines is found in the book *Power Over People* by Louise B. Young. Specific questions can be referred to Ms. Young at 755 Sheridan Road, Winnetka, Illinois.)

Where To Take Your Case

In most states there is at least one agency which has jurisdiction over the construction of new power plants and power lines. In some states the public utility commission has this responsibility. In others there may be a power siting authority or another environmental agency which must give a utility permission to build any major new facilities. Moreover, there are often state agencies which must approve a utility's plans for air and water pollution control facilities. An inquiry to your state's regulatory commission or environmental agency will help you determine exactly which agency, if any, has jurisdiction over new power facilities. Practically any government agency decision regarding such facilities can be appealed in court.

So far, decisions by state agencies to halt construction of new power facilities have been rare. But the environmental awareness of regulatory commissions is growing and, as more challenges are brought by citizens' groups, substantial victories are certain to emerge. In a precedent-setting decision in 1973, the Maryland Public Service Commission upheld environmentalists' contentions that one of two new plants proposed by the Potomac Electric Power Company was not yet needed.

Your own municipality or county may have laws which limit the construction of new power plants or power lines. Many citizens' groups which have brought cases before their county commissions have learned that these agencies are often dominated by business interests and will rubber-stamp almost any development proposal. However, some citizens' groups have built up enough pressure to convince their elected officials to turn down proposals for new power facilities. Some communities have passed legislation requiring high voltage lines to be placed underground. Others have discouraged the construction of new power plants by levying high property taxes for such facilities.

If your group wishes to oppose construction of a nuclear plant, you can take your case before the Nuclear Regulatory Commission. For each new nuclear plant, the NRC holds two sets of hearings: one for a construction permit and one for an operating license. It is important that citizen challenges begin at the construction permit stage—before substantial amounts of money have been invested by the utility. Thus far, the NRC has barred construction only in outrageous circumstances, such as the case of a geologic fault underlying the proposed plant site. Nevertheless, intervenors have caused substantial delays in the construction of many nuclear plants, often forcing utilities to install better safety and environmental protection devices than had been planned.

The Federal Power Commission has jurisdiction over the construction of hydroelectric and pumped storage projects, and so do many state agencies. One environmental group in New York State has successfully delayed Consolidated Edison's pumped-storage project on Storm King Mountain for over 10 years by making one legal challenge after another before state agencies, the FPC and the courts. Pumped-storage projects do not make electricity--they store it. Since the storage process wastes energy, citizens have argued against these facilities on energy conservation and environmental grounds. Environmentalists have permanently stopped some hydroelectric facilities and are currently opposing several more. It is important to recognize, however, that a hydroelectric project is often preferable to a fossil fuel plant.

Building Your Case

You already know why you oppose the construction of a new power facility. Now your job is to convince others that the project is undesirable and that it is possible to stop or change the utility's plans. Start with your neighbors: pass out leaflets describing the potential effects of the proposed power plant or power line. Enlist the support of experts on subjects such as air and water pollution; they can help you to document some of the problems you are concerned about. You may decide to set up meetings to discuss issues and strategies and to organize rallies to display public concern.

When you take your case before the appropriate government agency, all of your arguments should be well-documented. Explain in detail the expected environmental effects of the new facility. No doubt the utility will produce elaborate studies showing that the public will be protected from environmental hazards (e.g., tall stacks will disperse air pollution). Then you'll probably need an expert witness who can refute the company's arguments. Citizens' groups often perform their own detailed studies of air and water pollution, radiation hazards, power lines and so on. Some groups have found that a college professor, research firm or state agency has already studied these problems; by talking to experts in your community, you may find that much of your work has already been done. In many cases, the utility's own past performance record will provide a citizens' group with valuable evidence. Records of air pollution violations and fish kills should be available from appropriate state agencies. You should try to prove that further violations can be expected if the new facility is built.

If the design of a new power plant does not include adequate pollution controls, your group should be armed with facts on the availability of better equipment. Among the controls that many citizens' groups demand are 99.5% efficient electrostatic precipitators, sulfur dioxide scrubbers, cooling towers and nitrogen oxide controls.

Citizens often encounter some very difficult problems in opposing the construction of new power facilities. By stopping a new project in their neighborhood, one group of citizens might simply be giving the problem to someone else. Furthermore, delaying or halting the construction of generating plants or transmission lines may contribute to power shortages. Thus, it is essential that your group's motives not appear to be purely selfish. *You must provide the utility with a viable alternative to its proposed facility.*

Energy Conservation As An Alternative

If your problem is the routing of a power line through a park or a power plant proposed near a wildlife refuge, your alternative is simple: just find a more suitable location for the power facility. However, if the facility will cause intolerable environmental problems wherever it is located, you must find an alternative which will eliminate the need for the power plant or transmission line altogether. Therefore, you must suggest ways in which substantial amounts of energy could be conserved.

Questioning Load Projections. The first step is to examine the company's argument in favor of the facility. It will probably have a series of growth projections and assumptions supporting its contentions that the facility is needed. You can check these projections against the projections on file at the FPC in the *Reliability Council Report* which covers your company. Sometimes this comparison unearths substantial discrepancies in company load projections.

An electric power system must always have enough generating and transmission capacity to serve its expected annual peak load. In addition, utilities require a *reserve margin* for contingencies such as equipment failure, and high demands due to extreme weather conditions. The FPC recommends that utilities maintain about 15-20% more capacity than their expected peak load. Utilities usually opt for a 20% reserve margin. Environmentalists and consumers argue that such a high reserve margin causes higher rates and additional environmental problems. Be sure that your company's demand projections do not include *interruptible sales* (as opposed to *firm sales*). Interruptible customers can be cut off when system power demands approach a company's generating capacity.

Utility load projections are usually little more than extrapolations of past growth rates. Rarely do utilities consider the effects of rising prices on future demands for electricity. Nor do their forecasts reflect the growing consciousness of their customers about conserving energy. Another source of errant predictions is the failure to recognize trends in *appliance saturation.* Many utilities have experienced rapid demand growth due to the increased use of certain appliances, such as air conditioners. As more customers buy these appliances, the market becomes saturated and the rate of growth in demand is likely to decline. Saturation figures for major appliances should be available from the utility. Detailed appliance saturation figures for most utilities can be found in the annual statistical issue of *Merchandising Week* (available in business libraries). You may need to file interrogatories to obtain certain information from the utility.

Your group should try to determine the year in which the entire capacity of the new plant would be needed, based on the company's load projections with a minimal reserve margin (perhaps 15%). Then determine by how much the company's expected annual peak must be reduced so that the additional capacity would not be needed. For example, a 500 MW plant would not be needed if a forecasted 5,500 MW peak load were reduced to 5,000 MW. With the help of a competent economist or engineer, you should attempt to design an energy conservation program

which could achieve the necessary reduction by the year in question.

Lawsuits, Strikes and Protests

Citizens' groups can use lawsuits to force changes in their power company's operations. Successful suits have been filed against utilities for violating air and water pollution standards and for causing air pollution damage and killing fish. In addition, Detroit Edison and Georgia Power have been sued for millions of dollars on grounds of racial discrimination in their employment practices. A knowledge of appropriate federal, state and local laws will help you to determine which legal actions could be filed against your power company.

Low-income families are often burdened by utility policies such as service cut-offs and the requiring of large deposits and late payment fees. Increasing numbers of citizens' organizations are challenging these policies before utility commissions.

Utility workers are becoming increasingly dissatisfied with their employers, and the power industry has recently been plagued by increasing numbers of strikes. Many utility workers are faced with the same callous company attitude that is well-known to environmentalists and consumerists. Utility workers can be valuable allies to citizens' groups, and vice versa. Said one worker striking Cleveland Electric Illuminating in 1973, "If they run their nuclear plants the way they run their other ones, we are all in a lot of trouble." Customers of the Duke Power Company in The Carolinas are coming to the aid of striking miners who are attempting to unionize company-owned mines in Kentucky. In return, the United Mine Workers are opposing Duke Power's rate increase.

One drawback of many utility challenges is that they depend too much on experts like lawyers and economists and usually do not involve large numbers of people. Some groups, however, have demonstrated that significant victories can be won through public pressure without legal proceedings. In 1969, a grass-roots organization in Chicago called Campaign Against Pollution (CAP--now called Citizens' Action Program) began pressing the Commonwealth Edison Company to clean up its power plants. When Edison complained that it would be too expensive, CAP built up public pressure against the company. Members passed out leaflets asking people to deposit an attached penny into Edison's bank account to help pay for pollution control. CAP publicized its demands with massive demonstrations at Edison's offices and at the company's annual stockholders' meetings. Hundreds of public debates between utility officials and CAP leaders generated publicity for CAP's demands and bad press for the company. CAP's opposition to Commonwealth Edison's rate increase finally persuaded the company that it was too expensive *not* to clean up, so Edison switched to low-sulfur coal. The group also convinced the city council to pass a tough new air pollution ordinance.

CAP's actions caused a significant reduction in Edison's rate increase. The 200 citizens who went to the first Illinois Commerce Commission hearing on the rate increase found themselves unwelcome. They trudged up 19 flights of stairs (because the elevator was broken) to find a room with 80 seats—and 60 of them were reserved for company lawyers. But by the end of the proceedings, the commission had slashed much of the increase and made the rest contingent on Edison's pollution control efforts.

Community orgainzers have found that utilities are excellent targets for citizen action since each company is regulated and all local citizens are its customers. Competent research staffs can dig up important facts and publicize them through demonstrations, meetings and press conferences. Georgia Power Project researchers found that the Georgia Power Company's property tax valuations on file at the State Revenue Department differed from its figures in Form 1. They determined that the company was saving $5 million annually in property taxes by using two sets of figures. The Project's careful research has helped it to build a growing movement around utilities in Georgia. Similar campaigns are being organized in other states (e.g. Arkansas, California, Minnesota and Vermont). For more information on utility organizing, write Paul Booth (CAP), Midwest Academy, 600 W. Fullerton, Chicago, Illinois 60614, or The Georgia Power Project, P.O. Box 1856, Atlanta, Georgia, 30301.

LIST OF ORGANIZING RESOURCES

Fund-Raising:

The Grassroots Fund-Raising Book, Joan Flanagan, the Youth Project, 1000 Wisconsin Ave., NW, Washington, D.C. 20007 ($5.25)

Fund-Raising in the Public Interest, David Grub and David Zwick Public Citizen, Inc., Box 19404, Washington, D.C. 20036 ($4.50)

Massachusetts Fair Share, 364 Boylston St., Boston, Mass. 02116

Illinois Public Action Council, 59 E. Van Buren St., Chicago, Ill. 60605

Citizens Action League, 4133 Gilbert St., Oakland, Calif. 94611

Organizing Training:

Center for Urban Encounter, 3410 University Ave., SE, Minneapolis, Minn. 55414

Industrial Areas Foundation Training Institute, 12 E. Grand Ave., Chicago, Ill. 60611

The Institute, 3814 Ross Ave., Dallas, Texas 75204 (affiliated with ACORN)

Mid-Atlantic Center for Community Concern, 554 Bloomfield Ave., Bloomfield, N.J. 07003

The Midwest Academy, 600 W. Fullerton St., Chicago, Ill. 60614

National Training and Information Center, 121 W. Superior St., Chicago, Ill. 60610

New England Training Center for Community Organizers, 19 Davis St., Providence, R.I. 02908

Organize, Inc., 814 Mission St., San Francisco, Calif. 94103

Movement for a New Society, 4722 Baltimore Ave., Philadelphia, Pa. 19143 (inter-organizational consulting)

Utility Project Resources:

State Utility Activist List, limited to three states per order ($1.50 per state, $25.00 for profit-making businesses)

National Utility Experts List ($1.50)

Utility Action Guide (free)

(all available from EAF, Publications Department, 724 Dupont Circle Building, Washington, D.C. 20036)

HOW TO DEVELOP
A PRICE-ANDERSON FLIER

When I first heard about Price-Anderson, I asked three questions: How can it be Constitutional? Why hasn't Congress repealed it? And what can I do to work toward getting this bill repealed? I discovered that Price-Anderson is Constitutional because the Supreme Court has ruled that it is, that Congress has not repealed this act because the utility lobby is too powerful, and that I could, and did, write a flier which grass roots groups can distribute by the millions to property owners all over the United States. This flier can also be made into a public service announcement for television and radio.

A FLIER FOR GRASS ROOTS GROUPS

By Fred Wilcox

Here are some ideas for writing a Price-Anderson Flier. When homeowners throughout America discover the content of the Price-Anderson Act, they will demand its repeal—immediately.

I first heard about Price-Anderson at a demonstration against nuclear power when one of the speakers announced: "The way to end nuclear power in America is to force Congress to repeal the Price-Anderson Act." I returned home and began research into what has to be one of the most Machiavellian pieces of legislation of all times. And the more I read the more convinced I became that the speaker was right: *The Price-Anderson Act must be repealed.* Working toward this end should be high on the priority list of grass roots activists.

The Price-Anderson Act was passed in 1957 to encourage the development of atomic power in the United States by limiting the liability of a utility company. Because no private insurance company in the world would risk insuring nuclear power plants, Congress, wishing to give a boost to this "much needed industry," passed the P.A. Act which set the ceiling of liability for utilities at 560 million dollars. As an additional bonus to the utilities, Price-Anderson made sure that the taxpayer would pick up $435 million of this tab, while the remaining amount, $125 million, would be guaranteed by private insurers.* Should a catastrophe occur, the funds for paying utility lawyers to contest the claims of property owners or injured individuals would come out of this 560 million, as well as reimbursement for any loss which might incur to the utilities' property outside of the reactor. In short, the public is paying for the utilities' insurance. If a citizen is injured or loses his/her home because of a nuclear accident, the citizen must pay for a lawyer to dispute the claim against the utility! Like being mugged, and then paying for your attacker's defense council.

Shortly after the Price-Anderson Act was renewed by a majority of both houses, 329-61 in the House and 76-18 in the Senate, a U.S. District Court in Charlotte, N.C. declared the Act unconstitutional on the grounds that it would allow:

"The destruction of the property or the lives of those affected by nuclear catastrophe without reasonable

certainty that the victims would be reasonably compensated . . . The Act unreasonably and irrationally relieves the owners of the plants of financial responsibility for nuclear accidents and places the loss upon the people injured by such accidents who are by definition least able to stand such losses."*

But the District Court's ruling was overturned in 1978 by the U.S. Supreme Court which apparently did not see the cruel ironies of the act and held it to be constitutional.

Since the Price-Anderson Act has been approved by a large majority of both houses, extended until 1987, and sanctioned by the Supreme Court, what can opponents of nuclear power do to work for repeal on a grass roots level?

One possible tactic is through the initiative process. Had California's Proposition 15 been approved in 1976, for example:

"It would have established the most comprehensive and stringent nuclear safeguard standards in the nation. First, the owners of nuclear power plants would have become fully liable for the consequences of a nuclear accident. Utilities would have been required to waive the liability limitations established by the Price-Anderson Act. If any operating plant did not waive the liability limitation, it would have been required to reduce operating power levels to 60 percent capacity within one year, and 10 percent per year for each additional year that the provision had not been met. Thus, if an operating plant had not waived the liability limitation after six years, it would have not been permitted to operate. Additionally, no new plant could be constructed unless the operators agreed to waive any liability limitations."**

What I propose involves neither legal maneuvering, nor considerable sums of money. After I discovered that insur-

---

* John Berger, *Nuclear Power: The Unviable Option* (Ramparts Press, 1977), p. 146.

** Nader, Ralph and Abbotts, John, *The Menace of Atomic Energy* (Norton, 1977), p. 347.

---

* These figures have subsequently changed, but the taxpayer still pays approximately 100 million of the liability premium.

ance companies include a clause in their homeowners policies that excludes coverage in the event of a nuclear accident, I began asking around my community to see how many people were aware of the Price-Anderson Act and/or this particular clause in their homeowners policy. I found that *none* of my neighbors had ever heard of the Price-Anderson Act. All were very surprised to learn that their home would be a total loss in the event of a nuclear meltdown. Therefore, I decided to write a PRICE-ANDERSON FLIER which grass roots organizations can distribute.

PRICE ANDERSON FLIER

DEAR HOMEOWNER,
BEFORE YOU THROW THIS FLIER AWAY AS JUST ANOTHER PIECE OF JUNK MAIL, LET ME ASK YOU ONE QUESTION. HAVE YOU EXAMINED THE SMALL PRINT IN YOUR HOMEOWNERS INSURANCE POLICY LATELY? IF NOT, PLEASE TAKE TIME TO DO SO. YOU MAY BE IN FOR A SHOCK.
Look for a specific clause which states that in the event of a nuclear accident your insurance company is not liable for damages to your home or property. This could mean that one day you might be forced to leave your home— without reimbursement. Please don't think I'm trying to

frighten you. But I believe that it is the public's right to be informed about matters as crucial as this.

WHY WAS THE PRICE-ANDERSON ACT PASSED?

The Price-Anderson Act was passed in 1957 because no private insurance company would consider insuring nuclear power plants. Analysis of the safety issues involved had convinced private companies that the risk was simply too great.

IF PRIVATE INSURERS WILL NOT INSURE NUCLEAR POWER PLANTS, WHO DOES?

The utilities have devised a self insurance scheme whereby each nuclear plant will pay 5 million dollars liability in the event of an accident. But 100 million of the 560 liability insurance is paid for by the federal government, which means that you, the taxpayer, are paying to insure the utilities' property but not your own!

HOW DOES THE PRICE-ANDERSON ACT WORK?

The Price-Anderson Act sets a liability ceiling for the utilities in the event of a nuclear accident. This ceiling is 560 million dollars. Though this may seem like a considerable sum, it must be distributed on a pro-rated basis to

people living in a very large area, possibly as large as the state of Pennsylvania or California. Estimates of a major accident vary from 6 billion to 280 billion dollars.

WHAT DOES THIS MEAN IN TERMS OF YOUR LOSS?

It means that you could conceivably receive only pennies or at best dollars for your entire loss.

WHAT IF YOU SHOULD BECOME ILL WITH CANCER FROM THE EFFECTS OF A NUCLEAR ACCIDENT?

You would have to prove in court that your cancer was caused by the accident. The Price-Anderson Act protects the utilities from liability for your illness by placing a 20 year statute of limitations on damages. Many cancers take ten, fifteen, and even twenty years to develop.

SUMMARY

In the event of a nuclear accident, you will be covered neither by the federal government, nor by your private insurance company. You will be forced to leave your home, forced to argue for compensation in a court where a utility lawyer will dispute your claim. The utility lawyer's salary

will be paid for out of the 560 million liability fund. This means that your taxes will be paying for someone to dispute your claim for damages!

As a homeowner and father of four children, I would urge you to join me in calling for the repeal of the PRICE-ANDER-SON ACT. The 560 million liability ceiling is a cruel hoax, designed to protect utility companies from their responsibility to the American people. PRICE-ANDERSON is the insurance companies' admission that an accident could very well happen. PRICE-ANDERSON is evidence that the utilities not only fear an accident, but have made certain that when it happens the American taxpayer will be left holding the empty bag.

I URGE YOU TO WRITE, CALL, OR CABLE YOUR CONGRESSPERSON TODAY, DEMANDING THE IMME-DIATE REPEAL OF THE PRICE-ANDERSON ACT. I URGE YOU TO JOIN ME IN REFUSING TO PAY FOR YOUR UTILITY COMPANY'S INSURANCE. I URGE YOU TO JOIN ME IN DEMANDING THE REPEAL OF THE PRICE-ANDERSON ACT, BEFORE WE BECOME REFUGEES IN OUR OWN COUNTRY!

HOW TO USE THE PRICE-ANDERSON FLIER

The Price-Anderson Flier should be taken from door to door in your community. Take time to answer questions about the flier and the Price-Anderson Act. Before you begin distributing the flier it would be a good idea to do some role playing. Ask one another questions that might be asked when you distribute the flier. Be certain you know every aspect of the Price-Anderson Act,* and, if possible, take a copy of the Act with you.

* For a good description of Price-Anderson Act see: "Price-Anderson: A No-Fault Nightmare" by Senator Mike Gravel, in *Countdown To A Nuclear Moratorium,* (available from Environmental Action Foundation, 724 Dupont Circle Building, Washington, D.C. 20036, $5.00)

With the Price-Anderson Flier your organization could include a form letter calling for the repeal of the Price-Anderson Act and addressed to a Congressman or woman or Senator from your state. If you can afford it, include a pre-addressed envelope with the form letter. Encourage people to sign the letter and mail it to their representative as soon as possible. After a week or two has passed, it would be a good idea to follow up distribution of the fliers with a house to house canvass. Once again, answer questions and encourage people to sign the form letter(s).

UNVEILING OF THE PLAQUE ON THE WEST STANDS ON THE OCCASION OF THE FIFTH ANNIVERSARY, DECEMBER 2, 1947. . . Left to right are AEC Commissioners William W. Waymack and Robert F. Bacher, Farrington Daniels, Walter H. Zinn, Enrico Fermi, and R. M. Hutchins, Chancellor of the University of Chicago. (The West Stands were demolished in 1957, but the plaque remains at the site.)

Photo credit: Argonne National Laboratory

HOW TO REFUTE ARGUMENTS FOR NUCLEAR POWER

The arguments used by the utilities and government supporters of nuclear power are becoming increasingly hackneyed. As new studies are released which demonstrate that nuclear power is dangerous, expensive and utterly unnecessary for America's energy needs, it becomes easier to answer the arguments of nuclear proponents. The questions and answers in this section are not designed to make anyone an expert in the field of energy or nuclear power, but to offer those who have not read extensively about nuclear power the opportunity to see that there are answers to such absurd statements as "nuclear power will save America from its dependence on OPEC."

BASIC QUESTIONS AND ANSWERS ABOUT NUCLEAR POWER

By Fred Wilcox

Q: Isn't nuclear power a cheap way to generate electricity?

A: No, the dream of early developers of nuclear power to produce electricity which would be "too cheap to meter" has failed. In recent years capital costs for reactors have soared and the costs per installed kilowatt has gone from $243.3 in 1972 to $526.9 in 1977 (compared to $307.8 for coal plants). "According to energy economist Charles Komanoff these capital costs will continue to increase, reaching $2000 per Kw in 1986-7." (Nuclear Information & Resource Service fact sheet)

All this means that the consumer pays more for nuclear powered electricity than for coal or oil-fired plants. "Since 1972, nuclear-powered utilities received rate hikes more than 27 percent larger than those granted utilities with little or no nuclear investment." (Deborah Schoch, Critical Mass Journal, 1978)

Nuclear power is not cheap now and it will become even more expensive later as uranium deposits dwindle and the uranium cartel raises prices.

Q: Regardless of the costs of nuclear power, we must have it to meet our energy demands.

A: Nuclear power supplies between 12 and 14 percent of our electricity today and 3 to 4 percent of our overall energy. Studies of American's energy uses have shown that we waste fifty percent of our energy. Other studies have shown that the U.S. could derive 30 percent of its electricity from using the excess heat and steam from industrial plants—cogeneration. "The economic potential of cogeneration at existing plants, with a rate structure set up to encourage it, is equivalent to 208 nuclear plants." (Not Man Apart, January, 1979)

We now have 65 operating nuclear plants in the United States and another 90 planned or in various stages of construction. This means that using cogeneration we could have enough surplus energy to equal all 155 nuclear plants. We are risking our environment and our lives for a source of power we do not need.

Q: But what about the increasing demand for electricity?

A: Most utilities have more power today than they need. For example, the state of New York has a 10,000 Megawatt surplus, yet continues to construct nuclear plants. Even if the demand should suddenly soar we could easily meet that demand through cogeneration.

Q: Won't nuclear power help free the United States from its dependence on OPEC?

A: "Vince Taylor of the Union of Concerned Scientists has said that if we turned off every oil-fired electric plant in the U.S., Western Europe, and Japan tomorrow and built over 200 new nuclear plants in their place, the percentage of our oil that comes from abroad would drop from 65 percent to 60 percent."* We simply don't use that much oil for generating electric power, and building nuclear power plants has done nothing to reduce our use of oil in industry, the home and automobile. Only through conservation and changes in lifestyle can we begin to reduce our dependence on oil.

Q: But doesn't nuclear power provide jobs and, therefore, isn't it good for the economy?

A: Yes, while a nuclear power plant is being constructed it does provide jobs. But once the plant is finished, fewer than 100 professionals and paraprofessionals are needed to run a reactor. Per each dollar invested, electrical plants produce fewer jobs than practically any other industry. What happens in a community where a nuclear power plant is being constructed is similar to a gold rush. The long depressed economy suddenly picks up, and there is a rush to the area by skilled and semi-skilled workers who hope to "put something away for a rainy day." But once the plant is finished the unemployment rate rises again and people are dislocated.

---

* Not Man Apart, January, 1979.

Q: But isn't it true that nuclear plants bring long-term prosperity to a community by increasing the tax base by millions of dollars?

A: Yes, this is true. But in addition to increasing the tax base nuclear power plants increase the cancer rate among those who live or work near the plant. A community should consider the short-term gains from tax revenues against the long-term misery caused by increased cancer rates. Is it really worth it to have a new firehouse if the volunteers will die an early and painful death from radiation-induced cancer? Does it really matter if a community has a new library when the children who use the library will be ingesting plutonium and will suffer from the constant exposure to doses of low-level radiation?

Q: Critics of nuclear power say it's dangerous, but no one has ever died because of nuclear power.

A: This is one of the more cynical arguments the utilities use to convince the American public that nuclear power is necessary and risk free. It is not true. In 1961, for example, the government's experimental reactor near Idaho Falls went out of control and three men were killed. In the Southwest, Navaho Indians who have been mining uranium for Kerr-McGee have been dying in increasing numbers from the effects of breathing radon gas in poorly ventilated mines. In Harrisburg, Pennsylvania, a young couple is suing the utility that owns Three Mile Island for the death of their stillborn child. And in the shipyards at Portsmouth, New Hampshire, workers are dying from cancer at alarming rates. Nuclear power and weapons testing are killing large numbers of human beings throughout the world. Radiation researchers like John Gofman and Rosalie Bertell have clearly demonstrated that radiation causes cancer, and it is common knowledge (even the utilities do not deny it) that nuclear plants routinely release radiation into the water and air. In the next decade this cynical argument will be disproven, unfortunately, by the death of more and more human beings from radiation-induced cancer.

Q: O.K., nuclear power may be dangerous, but at least it is a proven source of electricty, unlike solar or alternative technologies.

A: This is a very interesting argument because nuclear power is *still* very much in the experimental stage. For example, the reactor at Three Mile Island was rushed into service *before* it had been thoroughly tested in order that the utility could secure a multi-million dollar tax break. The Fermi reactor which nearly melted down outside of Detroit was in the experimental stage because no one knew just how a breeder reactor would work. Today nuclear power plants operate only fifty percent of the time because the technology simply has not been perfected, and the emergency core cooling systems of all reactors have never been tested.

On the other hand, cogeneration has been used to generate electricity in European countries for many years. Some nations in Western Europe get as much as 25 percent of their electricity from industrial generation of electric power.

Q: Environmentalists are always complaining about nuclear power but they never come up with any alternatives.

A: Environmentalists and other people interested in the survival of our planet are constantly coming up with alternatives to nuclear power. It is really up to the government and big business to decide whether profit is more important than people.

1. Cogeneration. Cogeneration is a method of harnessing the energy which is wasted when the excess heat and steam from industrial plants is released into the air. In Europe this method of harnessing energy for the production of electricity has been used successfully for some time. In Oregon, a grass roots group (the Eugene Future Power Committee) defeated a nuke proposed for outside Eugene. In place of the proposed nuclear facility, the utility is using steam from a pulp mill to "cogenerate" electricity.

2. Dams. Once oil was a cheap fuel for producing electricity and the utilities decided to remove many of the turbines they had installed in small dams throughout the country. Studies have shown that just by replacing these turbines we could equal our current nuclear capacity.

3. Wind, water and solar: There are literally scores of books available today which explain how we can use our natural habitat to produce clean, safe and efficient energy. People everywhere are proving that small, decentralized systems of power do work. For example, a group of tenants on the lower east side of New York built a windmill on top of their building which produces electricity. But Con Edison, rather than encouraging these power entrepreneurs, sued the tenants. In this case the tenants won and Con Ed is required to buy the excess electrical supply generated by the windmill.

The question is not whether it is possible to use human ingenuity to solve our energy problems, but whether the utilities and the United States government are willing to encourage the development of small decentralized forms of energy.

HOW TO PREVENT THE COMING OF ARMAGEDDON

"It began to rain. Mrs. Nakamura kept her children under the umbrella. The drops grew abnormally large, and someone shouted, 'The Americans are dropping gasoline. They're going to set fire to us!' (This alarm stemmed from one of the theories being passed through the park as to why so much of Hiroshima had burned: it was that a single plane had sprayed gasoline on the city and then somehow set fire to it in one flashing moment.)"

–John Hersey, *Hiroshima*

I was five years old when the Enola Gay dropped Little Boy on Hiroshima . . . My father was in the Phillipines training for the invasion of Japan and when the bomb was dropped, the invasion was called off. Later, whenever Hiroshima was mentioned, people would assure me that it was a good thing the bomb was dropped. Otherwise my father might have been killed.

Since Hiroshima, America has been involved in two "limited" wars, one of which lasted over ten years. But today, for the first time in three decades, sentiment against nuclear war is building. As we become more aware of the dangers of commercial reactors we are re-minded once again of the awesome destructiveness of atomic war. And, with each success in stopping the proliferation of nuclear plants, we become more convinced that we can do something to prevent the total destruction of our planet.

What we do and how we do it must always be a personal choice, but should every com-mercial reactor be closed down tomorrow, we cannot return to despair, apathy or amnesia. We must keep working together to tell our government and those governments endeavoring to become atomic powers that the people of the world wish to live in peace, and that NO NUKES MEANS NO MORE BOMBS, NO MORE WARHEADS, NO MORE WAR.

SOME ADVICE FROM "IRREVY": AN IRREVERENT, ILLUSTRATED VIEW OF NUCLEAR POWER

By John Gofman

John Gofman shares the belief with a growing number of Americans that if we do not stop the arms race nuclear war is inevitable. But he has not lost hope for the future, and in this article he offers some suggestions on what we might do to save our planet.

Those who *know* the truth are not the same as those who *love* it.

Confucius

This weekend is the 33rd anniversary of the introduction of the era of nuclear blackmail. The nuclear bombing of Hiroshima and Nagasaki, far from marking the end of the nightmare caused by the Nazis in Germany and the Imperialists in Japan, was the beginning of a greater nightmare overhanging the future of humanity everywhere.

Nuclear blackmail has two aspects in this country. First: if you don't participate in the nuclear arms race, you will be incinerated or enslaved by an enemy. Second: if you don't embrace a civilian nuclear power program, you will supposedly suffer an economic disaster whose central feature will be a loss of jobs.

Like all blackmail, nuclear blackmail is a bad bargain for its victims.

In exchange for a source of extremely expensive energy for which no need exists, society is forced to participate in history's greatest crapshoot with respect to health and life—for both this and future generations. In a world-casino, we create an astronomical quantity of some of the most persistent poisons imaginable, and the game we are told to play is one of hope — hope that the engineers and scientists can figure out a way to contain all the radioactive garbage. If the scientists and engineers could succeed in this gigantic containment experiment, we would get some energy which we can readily get much more cheaply from benign sources. If the scientists and engineers continue to fail in containment of radioactivity, people will get a legacy of cancer, leukemia, and genetic injury which will continue to plague mankind for generations. As Dr. Chauncey Kepford has pointed out, the radon gas problem from milling uranium extends the injuries out to *billions* of years.

In the other part of the casino, the bargain is equally bad, but it is more difficult for the public to perceive it, thanks to a myth. The myth is that the only purpose of having nuclear weapons is to deter the use of similar weapons by an enemy. The lullaby goes like this:

> The deterrent will deter,
> And war will not occur!

Why the Nuclear Arms Race Has to End in Holocaust

Since so many Americans appear to believe that nuclear weapons will never be used, it is important for me to state why I think it is *nearly a certainty* that they will be used in the next 10 or 20 years, and used in a big way, intentionally. Unless, of course, people like you succeed in making the crucial difference. I offer my viewpoint as a former Associate Director of Lawrence Livermore Laboratory (1963-1969) where about half of the USA's nuclear weapons are designed. But nothing I say is classified.

Governments—for instance, the USA and USSR governments— commonly point to the absence of nuclear war in the past 30 years as *proof* that nuclear weapons successfully deter nuclear war. My view is that this will remain true ONLY UNTIL one side achieves the technical breakthroughs which give it a "first strike capability"—which is always defined as the ability to initiate and win a war while suffering only acceptable damage in return. In 1941, for instance, the Japanese evidently *thought* they had a first strike capability; nations do not attack others if they expect UNacceptable damage, according to the meaning of the word "unacceptable." However, since the Japanese were mistaken about having a first strike capability, the carnage on both sides was terrible; the USA ended the war with a technical breakthrough—the atomic bomb.

Technical breakthroughs continue, of course—a fact which undermines the key assumption of deterrence, namely the assumption that it is *impossible* for either superpower to achieve a technical breakthrough which would give it a first strike capability.

Every few months, unclassified news appears about advances in anti-submarine warfare, anti-satellite weapons, lasers, particle-beam weapons, anti-ballistic and anti-cruise missile weapons. Add to this a host of technical developments on both sides which are successfully kept *secret*, and I have to view the myth of eternal deterrence as groundless, irresponsible, and reckless "wishful thinking." Furthermore, achievement of a first strike capability is not exclusively a matter of technical breakthroughs. Later, we will examine what our rulers consider to be "acceptable damage."

I know our nuclear weaponeers. Their very highest priority is to prevent the Soviet weaponeers from developing and then using a first strike capability on us. I know that some of you refuse to impute evil intentions like that to the Soviet Union, but when we see how the Soviet rulers terrorize and thereby enslave their own people, I think it is foolish to think they would not enslave us too, if they could achieve a first strike capability. I wish to make it very clear that I am not referring to the Soviet peoples; I am referring to a small group of bullies at the top.

Some of you who don't know me, may be jumping to the conclusion that I support nuclear weapons. Nonsense. Like you, I regard nuclear weapons as an abomination which mankind must get rid of. It's a question of HOW. And it's a question involving not only Soviet, American, and European destiny; three or four *billion* souls can be easily enslaved if only ONE super-power remains on Earth with nuclear weapons.

American weaponeers, entrusted with the defense of this country, necessarily have to assume the worst about the intentions of the other side; any other assumption in their profession would represent gross negligence. This is an important point, sometimes overlooked because of our disgust with any preparations for genocide.

Conceiving a First Strike Means Using It

Now for a few moments, please put yourself into the shoes of the American weaponeers. In order to prevent the Soviet weaponeers from developing a first strike capability, our weaponeers have to develop defenses which can *nullify* the first strike power of new Soviet weapons. And clearly, the only way our weaponeers can develop defenses against a new weapon-system is first to dream up the new system for themselves. There is simply no time to wait to *see* what the Soviets have dreamed up, now that missiles can go from continent to continent, and from submarine to land, in a matter of minutes. So our weaponeers necessarily have to dream up first strike weapons themselves, if they are going to dream up our defenses against them.

Thus regardless of the pious claims that *this* country is not interested in going for a first strike capability itself, logic tells us that figuring out a first strike capability HAS to be our very TOP priority. This is logic, not inside information. As long as competing gangs of rulers are allowed to use *force* to resolve the conflicts they create in the first place, it is simply inevitable that the parties to conflict must necessarily seek a first strike capability, under the strictest military secrecy.

Most unfortunately, the logic proceeds one step more. What would happen when our weaponeers dream up a first strike capability for which they can dream up no defense? They would simply have to assume if THEY can do it, so can their counterparts in the Soviet Union. At this point, the survival instinct would almost certainly overwhelm competing moral considerations. In an act of *self-defense*, the USA would almost surely rush to USE its first strike, as the only way to prevent a similar system being used upon us at a slightly later date.

Of course, the odds are approximately even that the *Soviet* weaponeers will be the first ones to achieve the first strike capability. Or to *think* they have. A miscalculation by one set of weaponeers just guarantees utter devastation on ALL sides instead of on one side.

If you accept my premise that technical break-throughs in the next decade or two may indeed provide one superpower or the other with a first strike capability—and remember that such an achievement depends also on the rulers' secret standard of what constitutes ACCEPTABLE damage to their own peoples—then it follows that the intentional use of nuclear weapons is a near certainty. While experience in a nuclear weapons lab was not necessary to figure out this logic—and indeed, some others who have *not* had such experiences have reached similar conclusions—it is significant that my experience at the Livermore Lab is not able to provide me with ways to *fault* the hideous logic I have just outlined.

As I see it, no matter which side gets the winning dice in this area of the nuclear casino, the likelihood of intentional nuclear holocaust somewhere, and fairly soon, is far, far greater than the American public presently realizes.

Widget-Fixing and Worse: Unilateral Disarming

If you are tracking with me this far, then it will be clear why the secret research into new weaponry by both superpowers is where the action really is. And from such analysis, it follows that it is as MEANINGLESS for concerned citizens to fuss over counting old warheads, reducing "overkill," trimming percents off of military budgets, or ratifying new SALT agreements, as it was for concerned citizens to fuss over the improvement of a valve or filter-system on a particular power plant, instead of trying to STOP nuclear power. Neither the danger of nuclear weapons, nor the danger of nuclear power, can be reduced by widget-fixing. Therefore we *must* think big.

I believe for the United States to make *small* unilateral gestures in disarmament would be merely widgeteering, and that for the USA to take *large* unilateral steps in disarming would be tantamount to giving the USSR a first strike capability.

Though my position on this may be presently unpopular with you, and I know that many of you are pacifists, I do oppose unilateral disarming. I believe that the amount of trust the Soviet leaders deserve is zero. And I think that accepting the idea, "Better Red than dead," would be the same kind of error the Jews in Europe made, in their refusal to believe in the total barbarity of the Nazi Germans. I even believe that the Jews in the Warsaw ghetto who took a few Nazi monsters *with* them as they were killed, are more of an inspiration for the defense of human rights than people who let monsters commit their monstrosities with impunity. I realize that such a statement must grate badly on some of you, but since I have been listening to other points of view carefully, I ask only that you give mine fair consideration. We all share a very deep desire to solve the same problem.

I would like to add one man's opinion. It is only that— one man's opinion—but the man is the great Russian dissident and "father" of the Russian hydrogen bomb, Andrei Sakharov. Recently he met with Joan Baez in Moscow, and spoke into her recorder. Asked if he is a pacifist, Sakharov replied, "I am perhaps too old to be a pacifist. I have seen too much. . . You don't undersatnd this country. You don't understand this system, this government. Your country must stay strong."

What Difference Does It Make, Anyway?

I have been asked, in as much as I regard nuclear *holocaust* as so likely to occur, why I bother fighting nuclear power and other sources of cancer and genetic degradation. That's an important question.

"What difference does it make if they pollute the world, if they are going to blow it up anyway?" say some intelligent non-activists. My answer is that it makes *no* difference. It makes no sense to fight nuclear power unless we *also* stop the system which is going to give us nuclear war.

That fact has been troubling me profoundly since the very beginning of my involvement in the nuclear-power fight. But some of us are only "minimal bright," after all, and so it has taken me all this time to synthesize a concept which I think may be *helpful* and therefore worth discussing. . .

The Lust for Power As a Medical Disease

Although it is obvious that ordinary humans have only misery to gain from nuclear weapons and from nuclear power, it is equally obvious that governments globally are bent on acquiring more, not fewer, of these monstrosities. We simply must figure out *why* this is so, if we are going to stand a chance of reversing the momentum.

I have a simplistic view of the origins of the human dilemma, a sort of medical view, and some anthropologists and psychologists will disagree with it. Nevertheless, I shall share it with you as a basis for dialogue on this all-important issue.

Humans, like other animals, have the instincts of survival, which are enormously powerful instincts. Included in these survival instincts are acquisition of food, shelter, and freedom from predators. But whenever a function exists in an organism, almost invariably ERRORS also exist in the function. With the exclusion of accidents and trauma, this is what medicine is mostly about.

One major such error, and it seems to be peculiar to humans, is a *gross* distortion of the survival instinct. That distortion, pervasively evident in just about the entire recorded history of humans, leads to the desire in certain humans to acquire unlimited power over other humans and resources. Power, in this aberration, seems the only thing which makes such diseased individuals feel safe and secure about their own survival. While it appears that *most* social animals live in some sort of power hierarchies or pecking orders, the arrangement among *non*-human species does not threaten the whole species, and may even provide breeding and survival benefits for the species as a whole.

Not so among humans. Among humans, we observe cases of INSATIABLE appetites for power. We see the aberration in the form of humans who feel threatened, once they have accumulated half a billion dollars, unless they can acquire *another* half-billion dollars at the earliest possible moment. It may be that, to some individuals for whom taking control over everything on Earth is not enough, the space program is regarded as a glorious opportunity. These individuals may even sit up nights wondering if, out somewhere—sitting on the Moon and beyond—there might be some still independent ROCKS over which they could exercise their power!

This aberration of the survival instinct, though obviously manifest in the ecomomic sphere, is by no means limited to greed. Nevertheless, the *visibility* of greed over the centuries of recorded time has led many concerned humans to believe that the cure for this sickness would be the elimination of economic gain—for example, the abolition of capitalism. Of course, that direction was tried in Russia, but the lust for power is very much alive and thriving in the Soviet Union, even though personal wealth is not the objective of the rulers there. Wealth is only one way of having power over others.

That the Soviet's party-faithful, their scientific elite, engineers, and athletes are given some luxuries in the USSR is quite true, but that is really no different from the doling out of modest luxury by the power-structure in the non-communist world to scientists, physicians, lawyers, the military, and even to part of the laboring class, in the effort to buy loyalty. And it works, because these servants are satisfying their own insecurities about survival. If a person's obsession with gaining power is strong enough, he usually makes the perfect, and perfectly unprincipled, sycophant. Even though he may start at the lowest rank in the structure, he will usually manage to scramble and ingratiate himself into a spot far above his starting point. In our system, such climbers may be called "self-made men" when they hit the millionaire category.

The disease, whether in a capitalist or a so-called socialist economy, is a lust for power. Since it is obvious that power-lovers seldom if ever operate in the best interests of the vast bulk of humans, they have to take care that they keep the vast bulk of humans under control. A variety of ruses are employed for this purpose. Although their objective is power and control over humans—keeping them in their proper place—the methods are disguised not merely as "the way things ARE," but also as the way things HAVE to be, as a sort of natural *law*.

The Disease Leads to Nuclear Power

Control is the main attraction making nuclear power so popular with the power-privilege class of both the Soviet Union and the Western democracies. Every aspect of society is arranged, if possible, to centralize control, to guarantee the exacting of tribute from the peons of society. The control of energy supplies is just one of the manifestations of control in general. That is why decentralized solar power gets only lip-service from our rulers.

One very successful way to control peons is economic blackmail. Do as we want, or you will lose your job. In general, it is not presented as bald-faced blackmail, but rather as the dictate of an impersonal *economy*. It is a most effective ruse, for a *normal* survival instinct is present in people at all levels. And not too many people out of the total are more than a few paychecks away from the bread-line. There is no other way except economic blackmail to account for the scientists, the engineers, and the physicians, who know about the real hazards of nuclear power, for example, but do not speak out. Or, if they do speak out, they repeat such rubbish as, "A solution for managing radioactive poisons will be found," or another variant like, "A cure for cancer will be found." The panic of coping with their immediate survival requirement, which is the secure income, is so great that they construct a wall of rationalizations which prevents them from clear thinking about their own long-term welfare or that of their children, and *their* children.

As I see it, the ultimate reason that humanity finds itself witnessing a program to make people everywhere artificially dependent for energy upon some of the deadliest and most persistent poisons imaginable, is that the species has let control of society pass into the hands of its greediest, most power-obsessed, least principled members. And society lets the control *stay* there thanks to economic blackmail. When people themselves are polled about their own desires regarding energy, at least in this country, they overwhelmingly declare themselves in favor of solar power. But their desires do not count.

The imposition of nuclear power is one of the things I have in mind when I reach the following conclusion: The aberration of the survival instinct in certain humans is a most serious disease—far more serious for the species than cancer or heart disease or even the plague. For everyone here understands, I am sure, that acceptance of nuclear power means the acceptance of premeditated random murder by radiation, as a legitimate policy of modern so-called "civilization." And yet

the power-elite seems ready to stop at nothing to ram nuclear power down our throats. In service to the elite, power company hirelings even photograph license plates in order to identify people attending anti-nuclear rallies—which would certainly help the police to round them up at a later date, if ordered to do so.

Resisting with Civil Disobedience

In your efforts to help stop nuclear power, some of you have decided to commit non-violent civil disobedience. Laying your body on the line is a powerful statement and educational tool, as long as it enables you to multiply the number of people who thereby come to agree with you about the importance of your stand.

Nowadays, the media will cover your story, in case there is blood spilled on the ground, but unfortunately it will carry hardly a word about the substance of your cause. At the Seabrook rally, the media people almost never left their tent, where they were apparently waiting for action of the conflict-kind. I read reams of coverage about Seabrook afterwards, and there were not even three paragraphs reporting the substance of the giant protest, though plenty of substance was publicly discussed at the rally. To make your case comprehensible to the general public, you will have to continue the tireless person-to-person education which deserves the credit for bringing the movement as far as it *has* come.

I think the power of your action depends upon the extent to which you make the case that it is NUCLEAR POWER which is violating the law. How can YOU be violating the law, when you are trying to *prevent* a crime—namely the premeditated random murder which is committed by every nuclear power plant in the country. Indeed, the random murder starts even before the plant is built, because the mining and milling of uranium *start* the murder process—politely referred to as "health effects" by government regulatory agencies.

It is clear there is going to be no relief from random murder policies under the Carter Administration. The Nuclear Regulatory Commission's most recent Seabrook decision is hardly a decision to stop nuclear power development; it may not even stop the construction at Seabrook for more than a few weeks.

The fact is that long ago, the government teamed up with industry to perpetrate such a fraud and deception concerning the safety of nuclear power and the danger of ionizing radiation, that it makes Watergate look like "small potatoes" by comparison. The cover-ups have been so gross that even Peter Bradford, a newish member of the NRC, has stated (according to the Union of Concerned Scientists) that truth was one of the "first casualties" of the government's program to promote nuclear power, and that the truth has been undermined by what Bradford calls "silenced concerns, and rigged or suppressed studies." Suppression has been practiced both on engineering *and* biological considerations. Anyone within the Atomic Energy Commission who dared suggest that radiation caused cancer and leukemia was immediately excommunicated, and subjected to a determined campaign of vilification and slander. I can tell you that from personal experience. And that policy is alive and well today in the Department of Energy under the apparent guidance of Dr. James Liverman. Therefore it is right not to let the system divert much of our efforts into costly and futile hearings before *government* panels.

Challenging the "Right" to Issue Murder Permits

If we had a *justice* system in the United States, instead of a *legal* system, I doubt that there are enough jail cells to accommodate the deserving members of the atomic energy establishment, for their crimes committed against humanity. I am happy to note that the Union of Concerned Scientists has formalized the activists' call for criminal proceedings in these matters. Let me quote some recent testimony by Daniel Ford of the UCS before Congress:

> "The question of possible criminal activities on the part of officials entrusted with protecting the public safety must be resolved. As part of its background review of proposed nuclear licensing regulations, we therefore recommend that Congress request . . . the Criminal Division of the U.S. Department of Justice to carry out an investigation of the conduct of former AEC officials, now NRC officials, to determine any role they may have had in a nuclear safety cover-up."

That's how the legal system should be dealing with the *perpetrators.* Now, how should the system be dealing with the civil disobedients? In my opinion, the civil disobedients have to make the case that the legal system has no right to try them on the basis of criminal trespass. The sanctity of property is automatically forfeited if that property is being used to violate peoples' inalienable right to LIFE. Just because Congress authorized an agency to issue a license for a nuclear power plant to commit murder, does not make it lawful. Under the Nuremberg Principles, this country declared that individuals have a *duty* to consider principles which transcend obligations imposed by the state. Furthermore, there is no provision in our own Constitution empowering Congress to issue permits to murder people at random. That simply has to be challenged. *

There is a whole profession in this land known as the law profession. To be sure, most of the people in that profession serve to maintain the power structure, not to challenge its depredations. Nevertheless, there are *some* humans in that

---

* *Honicker vs. Hendrie,* a lawsuit which challenges the "right" of the NRC to commit premeditated random murder by licensing nuclear power plants, was filed September 6, 1978 in federal court. Joseph Hendrie is chairman of the NRC. Jeannine Honicker is asking that the U.S. District Court in Nashville require the NRC to revoke the licenses of all nuclear fuel-cycle facilities, because murder-licenses are unconstitutional. Initial testimony for Honicker was heard on October 2, 1978. Details are available from Joel Kachinsky, attorney for the plaintiff, FARM LEGAL, 156 Drakes Lane, Summertown, TN 38483.

profession, and I think only a few of them are doing their job. Why are there not SCORES of lawyers in this state loudly protesting the conviction of Alliance members for criminal trespass? Where is their concern for justice? It would be nice indeed to have their support in explaining to the public WHO really ought to be on trial, and for WHAT!

The Disease Leads to Nuclear Weapons. . . and Their Use

Nuclear weapons are *another* thing which justifies my earlier statement that aberration of the survival instinct in some humans is a more serious disease than cancer, and heart disease, and the plague.

In the quest for power-extension, *force* has been useful for those seeking such power-extension. The use of force creates the need for so-called "national security," and requires a willing supply of cannon-fodder. Lo and behold, people have been willing to *die* for one set of rulers, when the competing set appeared to be even worse. That was the case with many Americans during World War II, and as I have already said, some of us find ourselves thinking that the American set of rulers today is the *lesser* of two evils.

But we peons have never enjoyed a choice between war and PEACE, between domination and NO domination, between blackmail and NO blackmail. I think it is our job in the anti-nuclear movement to figure out a way finally to create this choice for mankind. If not now, when? If not us, who? I don't have the answers, and I speak here only because I think I have some of the right questions. The answers may come from *you.* Your ideas and your achievements, for instance in building alternative communication networks and alternative technologies, are formidable, and most encouraging about what is possible.

I hope we can all accelerate our thinking, because meanwhile the military and technical establishments, inside and outside universities, jointly proceed with the effort to learn how to FIGHT and to WIN nuclear wars for the power elites which they serve. With respect to achieving a first strike capability and using it, the time-schedule is closely related to their concept—voiced only through underlings, if at all—of what constitutes "acceptable damage."

When you read the writings of the military-technical establishments over the past couple of decades, you observe that they talk less and less about the human casualties of nuclear war, and more and more about how many years it would take to bring the American economy back to the 1970 level. Mr. Bassett of the Federal Preparedness Agency explained it on a recent NOVA television program about American civil defense in the event of a nuclear war. After describing how bureaucrats, and computers, and records would survive a nuclear war thanks to elaborate underground bunkers, built in nine regions of the country just for them (not for ordinary citizens), Mr. Bassett said:

> "This idea that we're going to be wiped out and nothing [more] can happen, and we might just as well give up if we have a bombing, is not right, because we *can* reconstitute Federal Government.

We could have a *viable* government. . . . Sure, maybe we'll be living at a lower level of economy, but we'd have enough people. Maybe we'd have to go back to a 1920 status of people and numbers and economy, maybe even earlier than that. But we've got enough people, and we've got enough people in the Federal Government who know how to do things, and who know their responsibilities, so that we could have a viable government again."

Mr. Bassett, who looked a bit overfed on the TV screen, has a wonderful view. Even by enduring a nuclear war, the public could not get government off its back. Out of the bunkers and into the rubble would crawl a healthy horde of Mr. Bassetts to tax the survivors, and to rule on who owns the viable parts of the new 1920 economy, and who owns the rubble.

Mr. Bassett's point of view should be no surprise, since his job is to please the power-holders at the top who have PROVEN that they could hardly care less about human suffering. When I say proven, I have in mind abominations like arranging economies so that about a billion people are malnourished and afflicted with unnecessary diseases; like always *resisting* measures to stop the poisoning of industrial and agricultural workers; and in Russia, the enslavement of the entire population through terror.

The Task Ahead: Disease Control

For people like you, who think there may be such a thing as morality and inalienable human rights, who think the life-experience is interesting and worthwhile, eliminating the twin nuclear problems is fundamentally a struggle between humans and the disease known as powerlust. . . an aberration, a sickness, in the normal survival instinct of some humans. Whether this severe sickness is genetic in origin, I do not know. It may be. But I really don't think it matters whether it is or not. Not all humans manifest this disease, and some manifest it only mildly. Even if some fraction of every human generation will have an aberrant survival instinct, there is JUST NO REASON to allow such individuals to inflict the consequences of their disease on the rest of mankind.

Throughout history, the MOST diseased humans have been *permitted* by all the others to get to the top of the heap. Today, this means that they have the power to impose both nuclear energy and nuclear weapons on the rest of the species. To me, it is absurd and almost unbelievable that the Soviet and American populations permit their rulers to possess nuclear toys which they can freely use to poison or incinerate millions or billions of people for no reason at all.

In the face of monumental public confusion, misconception, and resignation, it is self-defeating to focus on widget-fixing. The public must be helped to understand that the deterrent will NOT deter nuclear war, and that nuclear energy perpetrates premeditated random *murder*. Those insights are what activate us, and they will activate others too.

The task of the activists, as I see it, is to learn how to make it IMPOSSIBLE for the sickest members of the human species to control all the others. In other words, we must eradicate the disease of power-acquisition permanently from human society. It must be gone before *new* abominable technologies are developed. For instance, mind-control. For instance, human cloning. And it must be gone before the Western and Soviet rulers get around to *merging* their power, for their greater security and consolidated control.

If the human experiment is to have a future, the only hope I can see is eradication of that disease—power-acquisition by some humans over others. It is far more important to eradicate that dieseae than it ever was to eradicate bubonic plague, tuberculosis, coronary heart disease, or cancer. And it is essential to eradicate power-acquisition everywhere on Earth, not just in the United States.

Some Reasons for Optimism

It is very easy to be pessimistic about all that I have just described, on the grounds that there is no reason to believe that the eradication of the power-disease is possible. After all, it has been THE way of life for centuries and millenia, everywhere on Earth. But I have recently been feeling considerably more optimistic, because of observing one human trait which we have not yet discussed in this review—namely, a desire, respect, and striving for *justice*.

It is rather amazing that some humans yearn for justice, considering that it is so rare a commodity and considering especially that a quest for justice can, in the short term, run squarely into conflict with the basic instinct for survival.

Yet the evidence is there and unmistakable that the number of humans who have placed justice highest on their list is not negligible, and this has been so over centuries. Many with deep religious convictions have for a long time felt strongly about justice, and here I refer to individual convictions rather than the generally spotty performance in this regard among organized religions. There have been the conscientious objectors to war, reaching far back into history. There have been the tax-resisters, who regard the payment of taxes to governments which commit the vilest of atrocities (not the least of which is the sponsorship of nuclear power) as a clear form of enslavement.

There are those who participated in the civil rights movement actively and courageously, in numbers which finally became too large and too effective to ignore totally. Encouraging too were the growing numbers who made some sacrifices to stop the American atrocity called the Vietnam War.

And most recently, there is the very encouraging sign of a vital and growing anti-nuclear movement, a movement which really seems to appreciate what the real issues are.

Nevertheless, the numbers who think justice is THE worth-while human goal, who understand that power-systems and justice are never going to be compatible, are still small. That should not surprise us, when we remember that only about

100 years ago, overt slavery was practiced in this country; that 60 years ago, women were not even allowed to vote; that 20 years ago, the movies were still suggesting that *the American Indians* were the savages, rather than the whites who devastated them and their culture; and that until the last decade, this society treated its physically handicapped, mentally handicapped, and homosexual minorities like sub-humans.

What gives me hope about our enormous task is the *evidence* that insistence on human rights is growing among the American people.

And I think hostility toward those who violate human rights—for instance, by poisoning people—will grow very much more rapidly now that the corpses are appearing among the labor force, thanks to *chemical* poisons ruthlessly administered by job-givers, back when nuclear power was still just a gleam in Dr. Glenn Seaborg's eyes. Probably some of you saw the ABC-TV documentary on July 14 (1978) entitled "Asbestos: The Way to a Dusty Death," which revealed the collusive efforts to *cover up* and then to *ignore* the hazards of exposing workers to asbestos—with the result that perhaps more than a MILLION people are already condemned to a miserable, premature death. Such atrocities are not the exception; they are the norm of behavior by those with the power to use economic blackmail.

The Past Nine Years

About nine years ago, I realized that the story of asbestos was about to be repeated with radioactive poisons. There is not, nor will there ever be, a credible way to prevent the irreversible contamination of the Earth by radioactive poisons in a nuclear power economy. Nuclear power is simply incompatible with human health. That became obvious to me as a chemist and as a physician.

So in the early 1970's, I advocated that we should close down existing nuclear power plants and never build any more of them. Sometime later, Dr. Ralph Lapp—who has earned good consulting fees for teaching electric utility officials how to manage questions raised about nuclear power by the pesky public—honored me by stating that I was one nuclear critic who was "beyond the pale of reasonable communication." I have had compliments in my day, but never one so nice.

In the early 1970's, there were others who were concerned about nuclear power too. The focus of many of them, however, was on getting a specific nuclear plant constructed ELSEWHERE, just so that it is not near *me.* Efforts to get a murderous technology moved into the backyard of *someone more ignorant,* I found nauseating.

However, I do like a proposal put forth by John Lane and Jay Kinney. If we were to insist that, if any of Jimmy Carter's proposed nukes are actually built, they must be built right in the middle of cities (which is where the power will allegedly be needed)—and especially sited right next to state capitol buildings where so many legislators waffle and bray about nuclear safety—we might rapidly end urban

apathy on this issue, and expose the hypocrisy about nuclear plants as "good neighbors."

In the past nine years, there has been plenty of nonsense about being afraid to be labeled "anti-nuclear," with the resultant thrust of "We just want to make nuclear power SAFE," a message which is hard to distinguish from Mr. Megawatt's *own* claim. Nevertheless, that approach dominated the nuclear "safeguards" initiatives which went down to defeat two years ago.

In my opinion, one inadvertently fortifies the industry's *falsehood* that nuclear power CAN be safe and acceptable, by focusing on widget-fixing—getting stricter rules for radioactive transport, decent testing of the Emergency Core Cooling Systems, better evacuation plans in case of nuclear "incident," improved security against nuclear terrorists (which means our assistance in creating a police-state), stepped up monitoring to track radioactive poisons AFTER they have been released irretrievably into our environment, bigger and better studies on waste burial as if we should comtemplate *creating* more wastes to bury, and even getting different formats for letting citizens speak before nuclear licensing boards whose function is to *grant* more nuclear licenses! To me, upgrading the safety of existing and future nukes is good only if we submit to HAVING nukes, but I do not consent to having nukes at all.

Since it is the *totality* of the nuclear problems which tells us that nuclear power can never be acceptably safe, we surely *must* continue educating the public about them—but without falsely implying the problems are solvable. I realize that even the move-it-somewhere-else approach to nukes, in the early days of the movement, *did help* to start the process of public education, and the chain of events which led to the exposure of the lies concerning the cheapness and safety of nuclear power. And finally to an openly anti-nuclear movement.

The reason that I am heartened by the present anti-nuclear movement, of which the Trojan Decommissioning Alliance, the Abalone Alliance, the Crabshell, Palmetto, Clamshell, and several other alliances, are the cutting edge, is that you recognize nuclear power for what it is—a crime against humanity, with its built-in premeditated random murder of humans, even humans yet unborn, and a manifestation of injustice heaped upon humans by a social system worshiping at the shrine of power of *some* humans to do whatever they want to *other* humans. It is that association—nuclear power as a violation of justice—which is exciting and heartening. It is a big, big advance from widget-fixing.

From Widgets to Justice by Steps

To be sure, it is possible to educate new people only one step at a time, so it is fine that much effort has gone into teaching the public about the fantastic economic rip-off which nuclear power represents. And it is good that labor-people are being helped to see how nuclear power takes jobs away; it is useful to show that the reason we need to expand the economy, is because energy is *destroying* jobs so fast that an expanding economy is the only way to avoid massive unemployment! The mythology that we need to use more energy to make the economy healthy is being well demolished.

But *if* rip-offs were all that is at issue, the nuclear power controversy, and the exposure of the lies surrounding it, would not mean so much to me. It is the fact that a rapidly growing movement is being built of people who are concerned about justice *as a whole,* and the relationship of nuclear power to additional abominations, which is heartening.

I attended the Seabrook Rally on June 25 (1978). There were some 18,000 to 20,000 people there. Lest any of you be confused about the cancellation of a civil disobedience action in association with that event, let me assure you that Seabrook was *not* 18,000 or 20,000 people just there for a picnic. The thousands who sat in a hot, baking sun responding to talks hour after hour, clearly understood very well what the relationship is of nuclear power to human rights. You just could not have been at Seabrook without realizing that a truly meaningful anti-nuclear movement centering around justice is alive, flourishing, and growing at an astonishing pace.

The meaningful anti-nuclear movement is on the cutting edge of a bigger movement to eliminate the disease of power-acquisition, and to build a world based instead upon justice and inalienable rights. Obviously the numbers have to grow immensely. We face two main obstacles to increasing our numbers:

Two Obstacles to Progress

(1.) ECONOMIC FEAR, which is based upon the threat that you and others will lose your jobs unless you go along with any and all abuses the power-system wishes to perpetrate upon you, from vinyl chloride poisoning, asbestos poisoning, herbicide poisoning, right through a long list to radiation poisoning.

(2.) FEAR OF ATTACK by another country, a fear which relates to the absurd belief that *countries*, rather than a few specific individuals, are responsible for wars; this widely held belief blocks the kind of clear thinking which might prevent nuclear holocaust.

Government is certainly *not* going to eliminate either obstacle to increasing our numbers. For one thing, the system does *not want* to get rid of unemployment. That should be quite obvious, since the system thrives on using the fear of unemployment. Since the government serves the system, and since unemployment is our system's essential tool for controlling people, it escapes me how some of my liberal friends can expect government *ever* to be constructive in eliminating the fear of unemployment.

Countering Economic Blackmail

I think it is up to us to explore some innovative ideas to counter economic blackmail. Many of you have made a great beginning by personally disconnecting from the larger economic system. But since that does not reduce the effect of economic blackmail on the great bulk of the population, it does not directly help increase the size of the justice move-

ment. And quite obviously, "economic disconnect" does not provide a mechanism which can protect you, or your fellow humans, or your descendants, from victimization by nuclear war.

To counter economic blackmail, perhaps we need to give more thought to ways in which *we* could take care of people facing disemployment by the system, not by subsidizing their unemployment checks, but rather by *using* their capabilities. There must be desirable and desired services they can provide to people in the justice movement, in exchange for salaries or services provided to *them* by people in the justice movement.

If we assume that every reasonably healthy adult is able either to take care of his own needs, as can almost every other creature on Earth, or is able to exchange needed services with other humans, it is a deep mystery why we can't establish mini-economies based on affordable tools right in the middle of the bigger economy. The Amish community in Pennsylvania does do it successfully. After all, humans with hand-tools ARE CAPABLE of meeting their needs for food, shelter, clothing, and education—providing they are not prevented by sky-high property taxes from having the necessary land.

The paradox of capable humans who can not find work to do, is one which has apparently defied solution for many generations now. And not just in capitalist economies. Recently I read that Cuba has now developed an unemployment problem among its growing number of college graduates. Nevertheless, I think the unemployment paradox, and its concomitant economic blackmail, *can* be eliminated if they receive your clever attention.

With regard to employed people rustling up the money to employ people *who lose their jobs in the system,* I hope you will ponder upon the following figures. A report released last month by the White House's Office of Management and Budget states that Americans now spend about 700 million hours each year filling out federal forms, and that the estimated cost of preparing and processing all this paperwork is $100 billion a year. That's a reduction in useful income of about $500 for every man, woman and child in the nation, or $2,000 per family per year. But we don't "see" it because it is mostly hidden in the price of what we buy. Government IS inflation, not an agency seeking to stop inflation. Just by eliminating *federal* forms, ten employed families could afford to hire one UNemployed family *to do useful work* for a salary of $20,000 per year!

Reducing the Power of Government

Since government actions have always ended up serving the power-elite, I consider government to be a big part of our problem and *not* the solution. And by taking more than one-third of our incomes away in various taxes, government drastically reduces our already meagre capabilities. Every tax-dollar paid to government is a guarantee of one dollar less available for trying to SOLVE problems. No doubt my liberal friends will disagree again, but disagreement has no power to alter the truth.

Can you imagine, even in your wildest flight of imagination, how paying the salary of the head of the Department of Energy and his ten thousand employees is going to help *solve* anything? If you can believe that your taxes are helping to *solve* mankind's twin nuclear problems, then you can probably be blissfully narcotized into believing that SALT is a meaningful step toward preventing nuclear holocaust. I can not.

If government is not going to provide solutions, then it would seem logical that the sooner we have less government at all levels, the sooner we can free up some resources which *we* can direct toward solving problems. I would urge that you join, and help guide, the sentiment for tax-revolt in this country. There may be some unfamiliar bedfellows with you in that effort, but I would not assume that tax-revolt is either a racist backlash or a selfish backlash against the needy. For some, it surely is, but it is also an important backlash against tyranny. Virtual elimination of government would put enough money back in our hands so that we could directly, voluntarily, and warmly generous toward the people who *do need help.* The sense of *personal* responsibility for our fellow humans would be encouraged instead of killed.

Coping with "National Security" Concerns

When we consider the second obstacle to increasing the numbers in the justice movement—namely, the widespread concern about our "national security"—we face a very, very difficult task. Because as long as FORCE is tolerated by peoples everywhere as a way for rulers to settle their conflicts, then both military secrecy and the striving to achieve a first strike are inevitable.

I think the American people as a whole understand at the gut level that self-defense *is a necessity* as long as rulers have military force at their command, and therefore I think the American people will continue tuning out when they hear unilateral disarmament steps advocated. I think their gut feeling about an external threat is *correct.* *Your* gut feeling, that the deterrent will not deter and that nuclear weapons mean genocide, is ALSO *correct.* Mankind would have a better chance if there were communication, instead of confrontation, between the so-called militarists and the so-called peaceniks.

In science, when two apparently incompatible observations are each real, true, and valid, we know an explanation *has to exist.* We can not pick the data that we *like* the best, and throw the rest away! Reconciling conflicting data can be exquisitely difficult, but the effort almost always leads to new insights. With respect to the nuclear arms race, I have been trying to reconcile the valid views of the militarists with the valid views of the peaceniks. I think the medical model of the power-disease *does* explain why both are valid. And I think the explanation is of more than academic interest, because the way one analyses the driving forces in any situation—scientific or political—makes a very big difference in what you decide to *do* in terms of action. I think we have to cope with the following logic:

(1.) In order to eliminate *nuclear* weapons from this Earth, humans must also eliminate the use of force of ANY sort as an acceptable way to resolve conflicts. And,

(2.) In order to eliminate the use of force, humans must keep the power-lovers OUT of positions of power.

Even if it were somehow possible to get power-wielders, worldwide, to renounce *nuclear* weapons, as long as power-lovers remain in charge of the world, they will—because of the nature of their disease—continue to harbor the idea of using FORCE in order to increase their power . . . and so the old "R&D" (research and development) will just begin on things even worse than nuclear weapons. Like mind control. The problem is keeping the power-lovers OUT of power.

As I stated earlier, I believe interference with the nuclear capability of ONLY the United States would be an invitation to holocaust. I am certain *that* opinion rubs many of you the wrong way, but I urge you to think it through very carefully again. The possession of nuclear weapons certainly reflects the willingness to use them under the "right" circumstances, and we peons are not even offered a definition of "right" before genocide would be committed in our names. It is truly an abomination of the first rank in every way. Although the *possession* of nuclear weapons is certainly a violation of human rights, I think NOT possessing them at this time would result in an even greater violation of human rights, by inviting either nuclear incineration or Soviet enslavement of 3 billion additional people. Either we learn how to take military hardware and nuclear toys out of the hands of power elites EVERYWHERE, or we are getting nowhere on this problem.

Reaching Ordinary People Worldwide

Our toughest problem is that public enlightenment for this momentous change must be worldwide. It is essential to learn how to reach the Russian people, the Chinese people, and people everywhere. If our own government were truly committed to eliminating the use of force, we might find communication satellites and other effective technologies at the disposal of such an effort.

In this country, we still have the freedom to learn, and to speak freely to others. The Russians no longer have this opportunity, because the Soviet power-structure has adopted Hitler's concept of the Thousand Year Reich, replete with every available tool of body and mind control. But it is safe to assume that the Russian people share the same goals as humans anywhere else, and that they are victims of the Soviet power system, not its perpetrators.

Somehow we must create a way to get through to the Russian people and to help them join the effort to eliminate the disease of power-acquisition by some humans over others, just as we need to do here. How to be effective should occupy the best of your minds, for there is not too much time left.

It would be appropriate if today's nuclear weaponeers were in the forefront of this movement. They should be clamoring more loudly than anyone for steps to make their disgusting jobs unnecessary. But then, waiting for *their* leadership would be like waiting for the medical and legal professions to lead the fight against nuclear energy!

Quite in addition to the normal survival instinct which makes weaponeers act to *protect* their jobs, I can tell you something else which happens to weaponeers. They get carried away with the elegance of their scientific and technical insights, which are undeniably clever. *Often* I have heard them refer to a bomb-design as "neat," and even more often as "sweet." I have heard *that* even from J. Robert Oppenheimer, who opposed development of the hydrogen bomb. . . and who was persecuted heavily for his opposition.

Will the USA and USSR Merge?

Because nuclear holocaust is a near-certainty, for the reasons I have tried to present, it is possible that the rulers of the USA and USSR may come to regard this particular technology as too dangerous for *themselves.* If so, they could probably arrange for mutual nuclear disarmament quite easily. But I do not think they *would* get rid of nuclear weapons; it is far more likely that *their* solution to the problem would be to merge and to *consolidate* their nuclear forces in order to defend themselves from the rest of the world. As long as nuclear weapons can be made by the Third World nations, what chance is there of the USA and USSR giving them up? Merger of the super-powers might protect Americans from nuclear incineration, but merger would drag us into an even deeper pit. A society dominated by the terror of getting sent to the Soviet-style "gulag" prison system is truly the antithesis of justice and freedom, and the *consolidation* of power instead of the *elimination* of power would virtually guarantee that there could never be a triumph for the justice-seekers.

Since merger of the super-powers would be such a nauseating way to eliminate our risk of nuclear incineration, we must devise other approaches.

The Nuclear Problems on a Scale of Injustice

I am trying to say it as I see it. The way I see reality is not encouraging. It is difficult indeed to know what to do. But the fact that a problem is difficult, just does not make it attractive to go solve some non-problems instead! In the face of enormous problems, I feel that time spent thinking is *well* spent. Activity for activity's sake can turn out to be wheel-spinning or widget-fixing.

One can always deceive oneself that small steps leading nowhere will somehow arrive at the trickiest, most difficult destination imaginable, but let us *not* retreat into such self-deception. Let us try to distinguish between the small random steps which lead nowhere, and the small steps which *do* advance toward an important destination. If the human species is worth worrying about, and if it has a lot of enormous problems which share a common cause, let us always ask ourselves if our steps are zeroing in on that cause.

In closing, I must ask a question. Why do we consider nuclear weapons to be the worst kind of monstrosity? Yes, 70 million Americans could get wiped out practically overnight, plus some fallout victims and descendants later. That is obviously monstrous. But compared to WHAT? Robert McNamara, President of the World Bank, has tried to tell us (October, 1976) that there are 900 MILLION "severely deprived human beings struggling to survive in a set of squalid and degrading circumstances almost beyond the power of our sophisticated imaginations and privileged circumstances to conceive. . . . Malnutrition saps their energy, stunts their bodies, and shortens their lives. Illiteracy darkens their minds and forecloses their futures. Squalor and ugliness pollute and poison their surroundings."

Including all ages, 70 million people worldwide face starvation *every year*, in an annual holocaust few Americans even know about, and 400 to 600 million people on Earth have diets inadequate to sustain normal brain development. That is an even MORE monstrous injustice than nuclear war. Isn't it?

Getting killed is something we regard as an injustice, and we are mad about it, whether the killing is done to us *fast* with bombs, or *slowly* with nuclear pollution from civilian nuclear energy. But if we were to fight nuclear power and nuclear weapons only out of concern for our *own* survival, while ignoring the murderous abuse of hundreds of millions of our fellow humans in the Third World, we would essentially be saying that it is more important to save life for Americans than it is to save life for Latin Americans, Africans, or Asians. It is *not* more important. To worry about the survival and health of just unborn *American* generations would be only a small advance beyond the so-called concern of those anti-nuclear activists who faded from the fight for justice as soon as the nuke planned for *their* backyard was canceled!

What I find so encouraging about the recent anti-nuclear movement is that its activists—all of you—seem to see the anti-nuclear movement in its larger context, on the cutting edge of a far, far bigger movement for justice throughout the whole human species and toward the other species on this planet. It just happens that we are learning how to make our contribution to this movement via the nuclear issues.

It is in that context that I regard what you are doing as having *immense* importance for all humanity. NO NUKES!

By opposing nuclear power we are demanding clean energy. The Task Force Against Nuclear Pollution has been gathering thousands of signatures on their clean energy petition and presenting these signatures to Congress as a way of showing that the American people want to live in a clean environment. Grass Roots activists should write for copies of the petition and work within their communities to gather signatures. For copies of and more information about the petition write to: Task Force Against Nuclear Pollution, P.O. Box 1817, Washington, D.C. 20013.

A CLEAN ENERGY PETITION

1. I petition my representatives in government to sponsor and actively support legislation to: (1) foster wide use of solar — including wind — power NOW, and (2) phase out operation of nuclear power plants as quickly as possible.

Sign here	Name printed clearly here	Date

Your street address (where registered to vote)	City	State	Zip

2. I petition my representatives in government to sponsor and actively support legislation to: (1) foster wide use of solar — including wind — power NOW, and (2) phase out operation of nuclear power plants as quickly as possible.

Sign here	Name printed clearly here	Date

Your street address (where registered to vote)	City	State	Zip

3. I petition my representatives in government to sponsor and actively support legislation to: (1) foster wide use of solar — including wind — power NOW, and (2) phase out operation of nuclear power plants as quickly as possible.

Sign here	Name printed clearly here	Date

Your street address (where registered to vote)	City	State	Zip

4. I petition my representatives in government to sponsor and actively support legislation to: (1) foster wide use of solar — including wind — power NOW, and (2) phase out operation of nuclear power plants as quickly as possible.

Sign here	Name Printed clearly here	Date

Your street address (where registered to vote)	City	State	Zip

5. I petition my representatives in government to sponsor and actively support legislation to: (1) foster wide use of solar — including wind — power NOW, and (2) phase out operation of nuclear power plants as quickly as possible.

Sign here	Name printed clearly here	Date

Your street address (where registered to vote)	City	State	Zip

THE **TASK FORCE** IS HAPPY TO SUPPLY YOU with extra petitions, and you may print your own too For the names and addresses of helpers in your area, send a stamped self-addressed envelope to the **TASK FORCE.**

A Nationwide Petition Drive — Endorsed by RALPH NADER.

HOW IT WORKS — When you send signed petitions, *we sort them by Congressional District,* and we take them to the right Representatives Not just once, but again and again We *prove* that concern is growing in their own Districts IT WORKS! In areas where there are enough CLEAN ENERGY PETITION-signers to tip an election, elected officials at *all* levels are really paying attention THIS IS NO ORDINARY PETITION-DRIVE The **TASK FORCE** maintains a permanent registry of petition-signers — legible and computerized Signatures are never lost, they work for you repeatedly Through this nationwide petition-drive, you can WIN a sunshine future instead of a radioactive one by making nuclear and solar energy into major political issues Donations urgently needed for printing and postage

This is recycled paper.

PLEASE MAIL SIGNED PETITIONS TO: Task Force Against Nuclear Pollution, INC. Washington, D.C. 20013

INDEX
